市政与环境工程系列丛书

English on Municipal Engineering

市政工程专业英语

主 编 陈志强 温沁雪
主 审 赵庆良

哈爾濱工業大學出版社

内容提要

本书介绍了水处理和水资源方面的专业英语29篇，每篇又分为2部分，内容包括：水资源、水质、水质管理、给水处理技术、地表水污染、地下水污染、污水收集、污水处理传质理论、污水处理技术及微生物学等，信息量大，可读性强，使学生在提高英语能力和熟悉专业词汇的同时，又提高了专业技能。

本书可作为高等学校给水排水专业、水文水资源专业和环境工程专业本、专科生专业英语教材或课外专业英语阅读材料，也可为从事相关专业的工程技术人员获取相关的科技信息提供参考。

图书在版编目(CIP)数据

市政工程专业英语/陈志强主编. —2版. —哈尔滨：
哈尔滨工业大学出版社，2013.1(2023.1重印)
(市政与环境工程系列丛书)
ISBN 978-7-5603-1996-4

Ⅰ.市… Ⅱ.①陈… Ⅲ.市政工程-英语-高等学校-教材 Ⅳ.H31

中国版本图书馆CIP数据核字(2013)第004510号

责任编辑 贾学斌
出版发行 哈尔滨工业大学出版社
社　　址 哈尔滨市南岗区复华四道街10号 邮编150006
传　　真 0451-86414749
网　　址 http://hitpress.hit.edu.cn
印　　刷 哈尔滨圣铂印刷有限公司
开　　本 787 mm×1092 mm 1/16 印张14.5 字数300千字
版　　次 2013年1月第2版 2023年1月第4次印刷
书　　号 ISBN 978-7-5603-1996-4
定　　价 40.00元

再版前言

随着我国对外交往国际化进程的加快，涌现了大量国外的技术信息，通过专业外语针对性的学习，可以在较短的时间内，提高学生科技文体阅读技巧与应用英语的能力，了解国外技术发展动态。基于以上考虑，我们编写了市政工程专业英语教材，以期学生能够提高英语阅读能力和了解更多、更广的专业知识。

本书分为两部分，第一部分为水资源、水质和给水处理技术，主要包括水资源、水质、水质管理、供水及常用的给水处理技术等；第二部分为水污染与污水处理技术，主要包括地表水污染、地下水污染、污水收集、污水处理传质理论、污水处理技术及微生物学基础等。本书的特点是：(1)选材广泛，均为专业中所涉及到的最新内容和最常用的资料及数据；(2)难易适中，所选文体均经过处理，并能适应在校大学生实际英语水平；(3)专业性强，所选文体专业学术水平高，在安排文献章节时参照了专业课"水质科学与工程"的有关内容，可以使学生较早地了解专业知识；(4)信息量大，全书共 29 篇，每篇均有 2 部分阅读材料，可以使学生通过阅读获取与专业相关的科技信息。

此次再版由哈尔滨工业大学陈志强和温沁雪主编，赵庆良教授主审。感谢给排水专业部分同学在本书的编写中收集了大量的资料，感谢哈尔滨工业大学吕炳南教授、任南琪教授对我们的支持，感谢刘冬梅、荣宏伟、于忠臣等参与本书的编写、修改工作。在本书的编写过程中，参考了大量的 Internet 上的资源及相关参考书，在此对其作者表示衷心感谢。由于编者水平所限，疏漏及不妥之处在所难免，敬请读者批评指正，使本书在使用的过程中不断得到改进。

编　者

2013 年 1 月

Contents

PART ONE

PART TWO

PART ONE

Unit 1 Hydrologic Cycle

Water is the source of all life on earth. The distribution of water, however, is quite varied, many locations have plenty of it while others have very little. Water exists on earth as a solid (ice), liquid or gas (water vapor). Oceans, rivers, clouds, and rain, all of which contain water, are in a frequent state of change (surface water evaporates, cloud water precipitates, rainfall infiltrates the ground, etc.). However, the total amount of the earth's water does not change. The circulation and conservation of earth's water is called the "hydrologic cycle". The hydrologic cycle module has been organized into the following sections (see Fig. 1.1).

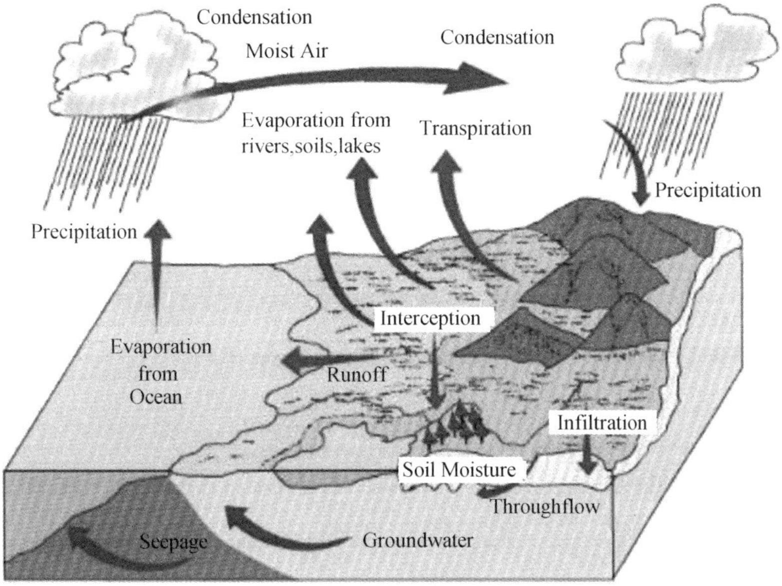

Fig. 1.1 The hydrologic cycle

The Earth's Water Budget

Water covers 70 percent of the earth's surface, but it is difficult to comprehend the total amount of water when we only see a small portion of it. The oceans contain 97.5 percent of the earth's water, land 2.4 percent, and the atmosphere holds less than 0.001 percent, which may seem surprising because water plays such an important role in weather. The annual precipitation for the earth is more than 30 times the atmosphere's total capacity to hold water. This fact indicates the rapid recycling of water that must occur between the earth's surface and the atmosphere.

Evaporation

Water is transferred from the surface to the atmosphere through evaporation, the process by which water changes from a liquid to a gas. Approximately 80 percent of all evaporation is from the oceans, with the remaining 20 percent coming from inland water and vegetation. Winds transport the evaporated water around the globe, influencing the humidity of the air throughout the world. Most evaporated water exists as a gas outside of clouds and evaporation is more intense in the presence of warmer temperatures.

Transpiration

Transpiration is the evaporation of water into the atmosphere from the leaves and stems of plants. Plants absorb soilwater through their roots and this water can originate from deep in the soil (For example, corn plants have roots that are 2.5 meters deep, while some desert plants have roots that extend 20 meters into the ground). Plants pump the water up from the soil to deliver nutrients to their leaves. This pumping is driven by the evaporation of water through small pores called "stomates", which are found on the undersides of leaves. Transpiration accounts for approximately 10 percent of all evaporating water.

Condensation

Condensation is the change of water from its gaseous form (water vapor) into liquid water. Condensation generally occurs in the atmosphere when warm air rises, cools and looses its capacity to hold water vapor. As a result, excess water vapor condenses to form cloud droplets. The upward motions that generate clouds can be produced by convection in unstable air, convergence associated with cyclones, lifting of air by fronts and lifting over elevated topography such as mountains.

Transport

In the hydrologic cycle, transport is the movement of water through the atmosphere, specifically from over the oceans to over land. Some of the earth's moisture transport is visible as clouds, which themselves consist of ice crystals and/or tiny water droplets. Clouds are propelled from one place to another by either the jet stream, surface-based circulations like land and sea breezes, or other mechanisms. However, a typical 1 kilometer thick cloud contains only enough water for a millimeter of rainfall, whereas the amount of moisture in the atmosphere is usually 10 ~ 50 times greater.

Most water is transported in the form of water vapor, which is actually the third most abundant gas in the atmosphere. Water vapor may be invisible to us, but not to satellites, which are capable of collecting data about the moisture content of the atmosphere. From this data, visualizations like this water vapor image are generated, providing a visual picture of moisture transport in the atmosphere.

Precipitation

Precipitation is the primary mechanism for transporting water from the atmosphere to the surface of the earth. There are several forms of precipitation, the most common of which for the United States is rain. Other forms of precipitation include: hail, snow, sleet, and freezing rain. A well developed extra - tropical cyclone could be responsible for the generation of any or all of these forms of precipitation. Amounts of precipitation also vary from year to year. In 1988, an intense drought gripped the Midwestern United States, disrupting agriculture because there was not enough rain to sustain crops. Five years later in 1993, the same area was subjected to severe flooding, greatly reducing the annual harvest because there was too much water for crops to grow.

Groundwater

Groundwater is all the water that has penetrated the earth's surface and is found in one of two soil layers. The one nearest the surface is the "zone of aeration", where gaps between soil are filled with both air and water. Below this layer is the "zone of saturation", where the gaps are filled with water. Under special circumstances, groundwater can even flow upward in artesian wells. The flow of groundwater is much slower than runoff, with speeds usually measured in centimeters per day, meters per year, or even centimeters per year.

Runoff

Runoff is the movement of landwater to the oceans, chiefly in the form of rivers, lakes, and streams. Runoff consists of precipitation that neither evaporates, transpires nor penetrates the surface to become groundwater. Even the smallest streams are connected to larger rivers that carry billions of gallons of water into oceans worldwide. Excess runoff can lead to flooding, which occurs when there is too much precipitation.

Summary of the Hydrologic Cycle

The hydrologic cycle begins with the evaporation of water from the surface of the ocean. As moist air is lifted, it cools and water vapor condenses to form clouds. Moisture is transported around the globe until it returns to the surface as precipitation. Once the water reaches the ground, one of two processes may occur: 1) some of the water may evaporate back into the atmosphere or 2) the water may penetrate the surface and become groundwater. Groundwater either seeps its way to into the oceans, rivers, and streams, or is released back into the atmosphere through transpiration. The balance of water that remains on the earth's surface is runoff, which empties into lakes, rivers and streams and is carried back to the oceans, where the cycle begins again.

Vocabulary

hydrologic	水文学的	stomate	气孔,气口
precipitate	沉淀,沉淀物	condensation	冷凝(作用),凝结(作用)
infiltrate	渗入,渗透	covergence	汇聚,聚合,合流
precipitation	降雨(量),降水(量)	front	锋面
evaporation	蒸发(作用)	topography	地形,地势
humidity	湿度,水分含量	breeze	微风
transpiration	蒸发(物),散发,蒸腾作用	penetrated	渗入,透入,渗透
nutrient	营养物,有营养的	runoff	径流

Reading Material A

The Water Balance

The water balance is an accounting of the inputs and outputs of water. The water balance of a place, whether it is an agricultural field, region, or continent, can be determined by calculating the input and output of water. The geographer C. W. Thornthwaite (1899 ~ 1963) pioneered the water balance approach to water resource analysis. He and his team used the water-balance methodology to assess water needs for irrigation and other water-related issues.

Every water transfer from a physical state to another and from one reservoir to another is known to be accompanied by an energy transfer, hence every water balance is linked with a similar temperature balance. In this page, only mass transfers will be considered.

Each earth system (the whole planet, atmosphere, a continent, an ocean, a region, a watershed) has its own water balance. The water balance of a system is the net gain of water deriving from the difference between the input and the output of water in a given period of time. The water balance can therefore be expressed by the following equation:

$$G = I - O$$

Where G is the net gain of water, I is the input, O is the output. All terms can be expressed as mass per time unit (e.g. tons per day) or volume per time unit (e.g. cubic kilometer per year) or height of water per time unit (e.g. cm of water per year). If G is positive ($I > O$), water accumulates in the system; if G is negative ($I < O$), water is lost from the system.

For a long-term balance or for closed system (e.g. the whole planet) is considered to be equal to 0.

Therefore, $I = O$.

The various processes of the hydrologic cycle are the input and the output terms of the water balance. It should be observed that each of these processes could be an input term for a system and an output term for another. For example, evaporation is an input for the water balance of atmosphere and an output for the water balance of a region; runoff is an output term for continent and an input term for oceans.

Global Water Balance

Considering all the oceans, which are the greatest reservoir of free water, total evaporation from their surface (output of water) is about 455,000 cubic kilometres in a year. At the same time, total evaporation from the soil, plants and water surfaces on continents is about 62,000 cubic kilometres. Therefore, total evaporation from all earth surfaces is 517,000 cubic kilometres. In a year the same amount of water will fall on the earth surface as solid or liquid precipitation (input of water), as water quantity on the earth is considered constant ($G = 0$). Precipitation doesn't occur uniformly on continents and oceans: 108,000 cubic kilometres fall on continent, 409,000 cubic kilometres on ocean surface. From the balance between evaporation and transpiration in each of the these two system (continents and oceans) it results that continents receive by precipitation about 46,000 cubic kilometres more water than by evaporation; this excess, commonly called runoff (down flow), flows on the continent or penetrates the underground reaching the sea in different ways and balancing the negative water budget of oceans (oceans lose about 46,000 cubic kilometres more water than that received by precipitation). This runoff is to be considered also the renewable available water, that the water we can consume yearly without spoiling the continents' water reserves.

The global water balance for continents is given by:

$$G = P - (E + R)$$

Where G is the net gain or loss, P is precipitation, E is evaporation, R is runoff. R is included in the output term.

The global water balance for oceans is given by:

$$G = (P + R) - E$$

Where the runoff is considered an input for the system.

In a long term, G can be considered equal to 0 in both the equations. Therefore,

$P = E + R$, that is to say: $108{,}000 = 62{,}000 + 46{,}000$ for continents;

$P + R = E$, that is to say: $409{,}000 + 46{,}000 = 455{,}000$ for oceans.

For the whole planet, the input sum should be equal to the output sum, so the runoff term is cancelled.

$$108{,}000 + 409{,}000 = 62{,}000 + 455{,}000 = 517{,}000$$

Water Balance in the Atmosphere

If we consider an air column, the water balance is determined by the following processes:

1. Water input due to the evaporation (E) from the surface underneath (both solid and liquid) or to plant transpiration in the column itself.

2. Water input or output due to atmospheric motions. The water mass crossing the column may have a bigger or smaller vapour content than thc column itself; therefore, it may add or take water from

the column. This is shown by the symbol C that is positive in case of addition and negative in the opposite case.

3. Water output due to precipitation (including frost and dew). The symbol is P and is negative because it diminishes the column water content.

Indicating with G the net gain of water vapour stored in the column of unitary atmosphere, with E the sum of evaporation and plant transpiration, with P precipitation and with C the net release of water due to air movements through the air column, the balance equation is given by:

$$G = E - P - C$$

Assuming a long term balance, the average annual net gain G is zero and therefore

$$E - P - C = 0, \text{ that is to say: } C = E - P$$

The average net release (input or output) of water vapour due to atmospheric motions is equal to the difference between evaporation and precipitation. It is positive, that is the column air will release water (in order to maintain the balance equal to 0), where evaporation is greater than precipitation. This normally occurs at low latitudes. It is negative, that is the column air will receive water from outside, where evaporation is lower than the precipitation. This normally occurs at high latitudes. These differences in gain or release of water in columns air over the latitude contribute to movement of humid air masses from equator to poles.

Vocabulary

water balance	水量平衡	accumulate	集聚
calculate	计算	region	区域
reservoir	水库,蓄水池	frost	霜,结霜
watershed	流域,水源区	dew	露,结露

Reading Material B

Protecting Water Resources

The availability and quality of water is a key factor in the nation's ability to protect public health, preserve ecological integrity, ensure sufficient agricultural production, and meet commercial needs. Issues impacting our nation's atmospheric water, surface water and groundwater (hereinafter referred to as "waters") must receive priority attention if the nation's public health and ecological and commercial objectives are to be met.

The engineering profession plays a key role in protecting and managing our nation's water resources. Engineers are leaders in developing the methods for ensuring that public health and ecological expectations for the country's water resources are satisfied.

Engineers are also the leading technical professionals in designing the infrastructure for managing water resources in areas such as flood control, water supply, water and wastewater treatment, channel and harbor construction and maintenance, and hydropower development.

Many national laws have been enacted that address water resource quality and management objectives. Among them are laws specifically oriented to water policy, such as the Federal Water Pollution Control Act, the Safe Drinking Water Act, and the Water Resources Development Act. Other federal laws, while not focused solely on water issues, also impact water policy. These include solid and hazardous waste disposal, fish and wildlife protection, and energy laws.

In 1972, Congress passed the Federal Water Pollution Control Act—commonly known as the Clean Water Act (CWA). The CWA is the primary national law for ensuring the "chemical, physical, and biological integrity of the nation's waters." The Act sets forth national goals and objectives and outlines the following activities to meet these goals: directs states to develop water quality standards for intrastate waters; develops effluent guidelines and best available technology (BAT) standards for industrial categories; establishes the National Pollutant Discharge Elimination System (NPDES) permit and enforcement program; establishes national technology-based limitations for municipal and industrial dischargers; regulates the discharge of dredge or fill material into navigable waters, including wetlands; assigns responsibilities to the states for non-point source pollution control; provides water quality-based controls for toxic pollutants; and establishes programs for funding the construction of wastewater treatment facilities.

The Safe Drinking Water Act (SDWA), first passed in 1974, strives to protect the quality of the nation's drinking water and provides limited protection of groundwater resources. The Act requires the following: issuance of BAT standards for tap water contaminants that threaten public health; monitoring and treatment of drinking water to meet federal standards; disinfection and filtration of water provided by public water systems; regulation of the disposal of wastes through underground injection; establishment of wellhead protection areas to protect wells from underground contamination; and establishment of programs for funding the construction of water treatment facilities.

The River and Harbor Act of 1899 and the Water Resources Development Act assign to the U.S. Army Corps of Engineers the responsibility for ensuring the navigability of the nation's rivers and harbors and providing flood protection. The Water Resources Development Act also provides funding for various water resource development projects that meet those objectives as well as for coastal erosion and levee construction projects.

Other national laws affecting our nation's water resources include the Resource Conservation and Recovery Act (RCRA), the Superfund Amendments and Reauthorization Act (SARA), the Coastal Zone Management Act (CZMA), and numerous fish and wildlife protection, federal land management, and energy laws. At present, procedures have been established through professional and occupational licensing laws and voluntary certification programs for ensuring the minimum competency of individuals engaged in water resource assessment, design, and management activities.

Vocabulary

water resource	水资源	discharge	排放(物)
ecological	生态的	dredge	挖掘，疏浚
infrastructure	基础	pollutant	污染物
hydropower	水力发电	disinfection	消毒，灭菌
set forth	宣布，阐明	filtration	过滤

Unit 2 Drinking Water Quality

The federal Safe Drinking Water Act, enacted in 1974 and amended in 1986, 1991, and 1996 is designed to ensure safe drinking water by establishing drinking water standards for treated potable water and by creating special protection for sources of drinking water, including groundwater sources like San Antonio's sole source aquifer.

Under changes made in the 1986 Safe Drinking Water Act amendments, the EPA required all public water systems to monitor for 16 inorganic (non-carbon compounds, such as nitrates, arsenic, fluoride, selenium) and 54 organic (any compound containing carbon) contaminants for which maximum contaminant levels have been established. The EPA currently has such levels in place for 81 contaminants, including total coliform, lead, and copper. In addition, the EPA requires monitoring of other organic chemicals for which levels have not yet established.

In Texas, the EPA has delegated authority for regulating drinking water to the state government. Currently, the state requires water systems to test for 126 chemicals, of which 73 have maximum contaminant levels. In addition, the state requires public water suppliers to test for bacteria and viruses and for secondary contaminants such as iron, manganese, and chloride that do not affect human health but lead to odor or taste problems.

Currently, the EPA has also begun a rule-making process to establish a standard for Cryptosporidium, a pathogen that can sometimes pass through disinfection and water treatment filtration to affect human health. In 1993 an outbreak of cryptosporidiosis—a disease caused by the pathogen—in Milwaukee, Wisconsin, led to 400,000 affected residents, including 4,000 hospitalizations and at least 50 deaths.

As part of the 1996 amendments, Congress, for the first time, funded the drinking water program by providing over $9.6 billion over a six-year period to states for improving drinking water infrastructure through the creation of state revolving funds, similar to the funding for wastewater treatment plants.

In Texas, the Texas Water Development Board is charged with overseeing the fund, which will provide low-interest loans to municipalities, utilities, irrigation districts, and other entities providing water for drinking. In FY 1997, Congress appropriated $740 million for direct and guaranteed loans, and Texas received a total of $71 million under the program. None of the funds can be dispersed without public participation in all aspects of the planning and implementation of the program.

Despite the passage and implementation of the Safe Drinking Water Act, drinking water in the United States and Texas is not always completely safe to drink. For example, in 1994, more than

45 million people in the United States were served by community drinking water systems that violated health-based requirements at least once. In Texas, violations are less common than nationally. According to EPA drinking water standards, Texas had 598 systems in violation, about six percent of all systems, affecting 1,658,406 people, or eight percent of the population, between 1994 and 1996. In 1997 only four percent of the population in Texas was served by public drinking water systems that did not meet health-based standards.

The Texas Natural Resource Conservation Commission implements the Safe Drinking Water Act for the state, and is charged with making sure that different types of public water systems regularly monitor their water supply, assessing the results, and enforcing the rules, including the maximum contaminant levels. In addition, under the 1996 amendments to the Safe Drinking Water Act, all states are required to develop a Source Water Assessment Program in which sources of regulated contaminants that could impact public water supply sources are identified. This assessment program will lead to a publicly accessible database of information on each water source and its actual and potential impacts of contamination sources. Currently, the TNRCC is implementing the program and the EPA is adopting rules. However, the first analysis from this assessment will not be due for several years.

Still, while most Texans get their water from municipal water systems which are afforded these basic protections, other residents of Texas receive their water supply from irrigation districts, individual water wells and increasingly from bottled water. Each of these sources have varying degrees of protection.

Vocabulary

drinking water quality	饮用水水质	lead	铅
potable	可饮用的	copper	铜
aquifer	含水层,蓄水层	manganese	锰
EPA	环保局(署)	cryptosporidium	隐孢子
arsenic	砷	pathogen	病原菌(体)
fluoride	氟化物	TNRCC (Texas Natural Resource Conservation Commission)	德克萨斯州自然资源保护委员会
selenium	硒	municipal	市政的
coliform	大肠菌		

Reading Material A

Existing Deficiencies in Water Supplies

The findings of the U.S. Public Health Service Community Water Supply Study in 1969 were a severe shock to laypersons and professionals alike. Approximately 17 percent of all public water supplies failed to meet one or more of the mandatory quality standards; 25 percent did not meet one or more of the recommended quality standards; more than 50 percent had major deficiencies in supply, storage, or distribution facilities; 90 percent failed to meet the bacteriological standards; and 90 percent had no cross-connection control programs. Conditions were particularly bad in the very small systems. Many were substandard in almost every respect.

The State of Washington, Division of Health, has accumulated some interesting statewide statistics in this regard. The Washington findings are similar to conditions prevailing in other states. It was learned that some 20 to 50 Washington communities, including some of the largest cities, are in need of additional filtration treatment. The exact needs are not known because of uncertainty that has arisen recently about the quality of supplies derived from so-called protected watersheds. The easy public access to any and all areas, including protected watersheds, that has been provided by the advent of all-terrain vehicles, trail bikes, and snowmobiles has made it very difficult or impossible to assure the safety of subsurface water supplies, which have no facilities for filtration treatment. Pending federal regulations that would require filtration of all surface water supplies appear to be a step in the right direction.

The study reported that 55 supplies, including all of the largest cities in the state, have open distribution reservoirs for storage of finished water. In the judgment of the Washington Division of Health, open distribution reservoirs are incompatible with present good public health practice, and they must now be covered under state regulations.

The Washington Division of Health is now requiring all cities to submit a comprehensive water system plan to identify present and ten-year future needs. It is anticipated that one-third of the cities already have most of the necessary data for such a plan; one-third will have a considerable amount of updating to do; and one-third have virtually none of the datas required for advanced planning, and will have to start from scratch. Two of the major unmet planning needs are coordinated planning among neighboring water utilities and minimization of the number of the new systems being constructed.

Statistics compiled by the State of Washington also reveal how little effort is being expended in the surveillance of water supplies by local, state, and federal regulatory agencies, as compared to similar efforts in air and stream pollution control. Expenditures for air pollution control were 10 times that for water supply monitoring and 200 times greater than for stream pollution control. The Washington study concluded that the incidence of disease is no longer an acceptable measure of the adequacy of public water supply systems, and there is a need to make this clear to the public, the press, and even the waterworks industry.

In addition to the deficiencies already mentioned with respect to public health, there are aesthetic considerations that deserve mention because of their importance to the consumer. Many water supplies contain sufficient iron and manganese to cause staining of laundry and plumbing fixtures, or even of building facings within the range of lawn sprinklers. Some waters are too hard to be satisfactory for many uses, or have corrosive or scaling tendencies that adversely affect plumbing in homes or factories. One of the most common complains about water quality concerns taste and odor. Bad-tasting water drives many consumers to other source of drinking water, such as private wells, cisterns, or bottled water. More often than not, such alternate sources of drinking water are of questionable bacteriological quality.

Vocabulary

deficiency	不足,缺陷	aesthetic	美学的
layperson	非专业人士	plumbing	室内管道(系统)
mandatory	强制的	hard	硬的
cross-connection	交叉连接	corrosive	腐蚀的,腐蚀物
bacteriological	细菌学的		

Reading Material B

Quality of Ground Water

For the Nation as a whole, the chemical and biological character of ground water is acceptable for most uses. The quality of ground water in some parts of the country, particularly shallow ground water, is changing as a result of human activities. Ground water is less susceptible to bacterial pollution than surface water because the soil and rocks through which ground water flows screen out most of the bacteria. Bacteria, however, occasionally find their way into ground water, sometimes in dangerously high concentrations. But freedom from bacterial pollution alone does not mean that the water is fit to drink. Many unseen dissolved mineral and organic constituents are present in ground water in various concentrations. Most are harmless or even beneficial; though occurring infrequently, others are harmful, and a few may be highly toxic.

Water is a solvent and dissolves minerals from the rocks with which it comes in contact. Ground water may contain dissolved minerals and gases that give it the tangy taste enjoyed by many people. Without these minerals and gases, the water would taste flat. The most common dissolved mineral substances are sodium, calcium, magnesium, potassium, chloride, bicarbonate, and sulfate. In water chemistry, these substances are called common constituents.

Water typically is not considered desirable for drinking if the quantity of dissolved minerals exceeds 1,000 mg/L (milligrams per liter). Water with a few thousand milligrams per liter of dissolved minerals is classed as slightly saline, but it is sometimes used in areas where less-mineralized water is not available. Water from some wells and springs contains very large concentrations of dissolved minerals and cannot be tolerated by humans and other animals or plants. Many parts of the Nation are underlain at depth by highly saline ground water that has only very limited uses.

Dissolved mineral constituents can be hazardous to animals or plants in large concentrations; for example, too much sodium in the water may be harmful to people who have heart trouble. Boron is a mineral that is good for plants in small amounts, but is toxic to some plants in only slightly larger concentrations.

Water that contains a lot of calcium and magnesium is said to be hard. The hardness of water is expressed in terms of the amount of calcium carbonate—the principal constituent of limestone—or equivalent minerals that would be formed if the water were evaporated. Water is considered soft if it contains 0 to 60 mg/L of hardness, moderately hard from 61 to 120 mg/L, hard between 121 and 180 mg/L, and very hard if more than 180 mg/L. Very hard water is not desirable for many domestic uses; it will leave a scaly deposit on the inside of pipes, boilers, and tanks. Hard water can be softened at a fairly reasonable cost, but it is not always desirable to remove all the minerals that make water hard. Extremely soft water is likely to corrode metals, although it is preferred for laundering, dishwashing, and bathing.

Ground water, especially if the water is acidic, in many places contains excessive amounts of iron. Iron causes reddish stains on plumbing fixtures and clothing. Like hardness, excessive iron content can be reduced by treatment. A test of the acidity of water is pH, which is a measure of the hydrogen-ion concentration. The pH scale ranges from 0 to 14. A pH of 7 indicates neutral water; greater than 7, the water is basic; less than 7, it is acidic. A one unit change in pH represents a 10-fold difference in hydrogen-ion concentration. For example, water with a pH of 6 has 10 times more hydrogen-ions than water with a pH of 7. Water that is basic can form scale; acidic water can corrode. According to U.S. Environmental Protection Agency criteria, water for domestic use should have a pH between 5.5 and 9.

In recent years, the growth of industry, technology, population, and water use has increased the stress upon both our land and water resources. Locally, the quality of ground water has been degraded. Municipal and industrial wastes and chemical fertilizers, herbicides, and pesticides not properly contained have entered the soil, infiltrated some aquifers, and degraded the ground-water quality. Other pollution problems include sewer leakage, faulty septic-tank operation, and landfill leachate. In some coastal areas, intensive pumping of fresh ground water has caused salt water to intrude into fresh-water aquifers.

In recognition of the potential for pollution, biological and chemical analyses are made routinely on municipal and industrial water supplies. Federal, State, and local agencies are taking steps to increase water-quality monitoring. Analytical techniques have been refined so that early warning can be given, and plans can be implemented to mitigate or prevent water-quality hazards.

Vocabulary

solvent	溶剂	hydrogen	氢,氢气
sodium	钠	acidity	酸度
calcium	钙	scale	水垢
potassium	钾	fertilizer	肥料
chloride	氯化物	herbicide	除草剂
bicarbonate	碳酸氢盐	pesticide	杀虫剂
sulfate	硫酸盐	landfill leachate	(垃圾)填埋沥出液
boron	硼	mitigate	减轻,缓和

Unit 3 Water Analysis

Physical, chemical, biological and radioactive variables vary widely in all raw surface waters and some high concentrations may be difficult to reduce during the treatment process. Public water supply standards vary between countries in amounts or concentrations of variables permitted and in the number for which tests must be made, for example, the Environmental Protection Agency of the USA (US EPA) lists some 139 variables.

Physical Variables

Turbidity and colour are the most common physical variables that need to be addressed during the treatment process. The levels of each of these have a large influence on the treatment to be applied and so it is important to have reliable measurements for both variables in raw water throughout the year. The necessary coagulant dose is directly related to the amount of turbidity or colour to be removed. When the levels of these variables are always low, more economical treatment can be used; if they are always high a more complex and expensive plant and process must be designed; and if they vary widely, which is common, the treatment plant must be sufficiently flexible to treat the extreme conditions as economically as possible.

The design of treatment plants must pay attention to several points when colour or turbidity of the raw water are high. The plant must have a large capacity for coagulant storage, supply and dosing; very good initial mixing of the coagulant and raw water must be achieved; and other factors must be considered relating specifically to colour and turbidity as described below.

Biological Variables

The common biological variables of relevance to the treatment process are bacteria, viruses and algae. All are present in surface waters but their numbers depend on conditions in the drainage basin. Rivers in regions with large populations and industrialized areas may be highly polluted, carrying large quantities of bacteria and viruses, while most pristine streams in sparsely settled areas of the world are relatively uncontaminated and have low numbers of bacteria and viruses.

Biological variables are much more difficult and complex to monitor than most physical variables. Identification and counts of organisms demand more training and a higher level of personnel expertise than for the simple reading of turbidimeters and colorimeters. Quantification of bacteria, and in particular viruses, require more complex equipment. Reasonably large water departments should therefore have equipment and a trained microbiologist or biologist to monitor bacteria and algae. Monitoring of viruses is more difficult but probably unnecessary for most water sources.

The number of bacteria and viruses are reduced during the treatment process in close proportion to the reduction of turbidity. If treatment is reducing turbidity by 95 percent, it may be assumed that bacterial and viral loadings are being similarly reduced. When turbidity of the filtered water is slight, subsequent sterilization is very effective and should consistently eliminate bacteria and most viruses provided the proper dose of sterilization is applied.

From the point of view of treatment operations, by far the most important organisms are algae. These organisms can cause serious problems in treatment plant basins by the accumulation of growths on the walls, efficient clogging of filters and causing taste and odour problems. It is therefore very important to perform algal identification and counts.

Chemical Variables and Contamination

Chemical variables are by far the broadest group to be identified and monitored in relation to design and control of the treatment plant and process.

pH The pH of the water indicates the degree of its acidity or alkalinity, reflecting the characteristics of the watershed or the underground rock strata through which the raw water has passed. Where limestone is predominant the waters are of high alkalinity and hardness, and have a high pH; conversely if limestone is absent the waters are lower in alkalinity, generally soft, and have a lower pH. The pH has a significant influence on the reaction of coagulants with the raw water. For maximum effectiveness of alum as a coagulant, the pH range is quite narrow, while that for ferric sulphate is wide. In treatment plants using alum as the coagulant (the great majority), the optimum pH is particularly important, or coagulant is wasted.

Measurements of pH should be made routinely (and recorded) for raw water, settled water, filtered water and water discharged to the distribution system. Samples should also be taken in the distribution system to monitor and pinpoint any changes.

Alkalinity Alkalinity must be present in raw water for coagulation to proceed, and for a satisfactory amount of floc to form. Its origin can be natural, having dissolved from alkaline rock in the watershed, or it may have to be added because some waters are of naturally low alkalinity. The most common coagulant in treatment throughout the world is "alum" $Al_2(SO_4)_3$.

Alkalinity in raw water is sufficient for the coagulation reaction to proceed in most cases. The design of the treatment plant must take alkalinity into account. In addition, alkalinity must be monitored during the treatment process. If the alkalinity is too low, provision must be made for addition, whereas if it is too high, the water will be hard and softening may be necessary.

Iron and Manganese Iron and manganese both cause problems in water supplies. Iron is more common and occurs in silicates from igneous rocks that are widely distributed across the world, whereas manganese is found more often in metamorphic and sedimentary rocks. Iron and manganese problems occur in lakes and reservoirs where anaerobic conditions reduce Fe^{3+} and Mn^{4+} to the soluble ferrous (Fe^{2+}) and manganous (Mn^{2+}) forms. In a water body that stratifies during warm, calm conditions, soluble iron and manganese may be trapped in the bottom anaerobic water layer. When the water column becomes thoroughly mixed by wind action and changes in water temperature the soluble iron and manganese become mixed throughout the water column and cause problems for the treatment plant, due to increased levels of colour, turbidity and organic matter.

Tastes and Odours Tastes and odours are quite common in water supplies everywhere because they are caused by a wide variety of substances, many of which readily enter water systems. Naturally occurring tastes and odours are often attributable to algae and blue-green algae. The chemical compounds which are most frequently responsible for incidence of tastes and odours include formaldehyde, phenols, refinery hydrocarbons, petrochemical wastes, acetophenone, ether and other contaminants produced by petrochemical industries.

Vocabulary

radioactive	放射性的	filter	滤池,过滤器
turbidity	浊度	alkalinity	碱度,碱性
turbidimeter	浊度仪	alum	铝
colour	色度	igneous	熔融的
coagulant	絮凝剂	formaldehyde	甲醛
dose	剂量	phenols	苯酚
raw water	原水	refinery hydrocarbons	精炼厂碳氢化合物
algae	藻类	petrochemical	石化的,石化产品
clog	阻塞	acetophenone	苯乙酮

Reading Material A

Water Problems

Some rain water which falls on the Earth is evaporated by the sun's heat. Some of it sinks into the ground. It may be used up by thirsty plants. It may reach a well or a spring. Most of the water goes back to the rivers, seas and oceans. This process then starts all over again. The process is called the Water Cycle.

The environment in which modern man lives is the result of a complex process which began with the origins of life on earth and, but for the presence of water, could not have developed as we now know it. All living things require water to survive, from desert plants that germinate and flower briefly after infrequent rain, to man himself whose dependence on water to drink and to produce food is fundamental. When we consider man's needs in the environment he has created for himself, the problems of water quality assume alarming proportions.

Primitive man realized very early in his development that water was essential to his existence; he needed water to drink, the plants and animals he required for food flourished only where there was an adequate water supply. He found that water could be useful as a defence against his enemies, that it was a useful source of power, and that too much could lead to chaos and destruction. In modern times, the horrors of drought in Ethiopia and the results of severe flooding in Bangladesh are vivid illustrations of the important place of water in the environment.

For thousands of years man has been aware of his need of water in the right quantity. But, as his technology developed, as urbanization began, so pollution commenced, and slowly the need was realized for water of adequate quality. It is a sad fact that in the industrialized societies of the world, greed and technology have outstripped science. This has resulted in many rivers and water courses becoming heavily polluted by urban and industrial effluents, and the population of towns and villages dependent on a river for water becoming exposed to epidemics of water-borne diseases and to excessive amounts of trace metals because of the lack of knowledge and equipment of monitor and control the quality of water in the environment.

This lack of knowledge is best illustrated by the fact that the science of bacteriology was not fully appreciated until the mid-nineteenth century, and for several decades after this many water supplies continued to be drawn from rivers downstream of effluent discharges which received little or no treatment, annual rounds of typhoid and cholera were a regular feature of life and death. Control of water quality by careful selection of sources and treatment processes has virtually eradicated water-borne epidemics in the developed countries. But there still remains a problem in the third world.

The lack of equipment is shown in the current concern over lead in the environment. It is only in the last 20 ~ 30 years that reliable equipment has been readily available to measure accurately lead (and other elements) in very low quantities and, more important, to assess the effect of lead on the human body.

Today the human race appears to have reached the point at which it has realized many of its mistakes in the misuse of water, and is laying down standards, recommendations and guidelines, to control water quality at various stages in the water cycle in an attempt to reverse a trend which, in developed countries, had almost reached disaster level.

The fraction of the water cycle we are concerned about here is the 3 percent existing as fresh water, and it is this 3 percent that is generally available to man for purposes of health, food (agriculture and freshwater fisheries), industry and leisure. Because of this, measuring, monitoring and controlling quality in this portion of the cycle is important to all societies.

When considering how to do this in detail, it is important to consider what happens in nature to affect the quality of water (contamination) and what man has superimposed on this (pollution).

Vocabulary

environment	环境	typhoid	伤寒
germinate	发芽,萌芽	cholera	霍乱
flourish	繁荣,茂盛	lay down	制定
chaos	混乱	recommendation	建议,劝告
epidemic	流行病	reverse	扭转
trace metal	微量金属	superimpose	添加,附加

Reading Material B

Surface Water Quality in Texas

Texas has approximately 305, 965 km (191, 228) miles of streams and rivers, of which 64,310.4 km (40, 194 miles) (21 percent) are considered perennial (having sustained flow throughout the year); nearly 2.63 hm^2(6.5 million acres) of inland wetlands and 0.69 hm^2(1.7 million acres) of coastal wetlands; more than three million kilometers of reservoirs and 5,155.9 km^2 (1,990.7 square miles) of bays; and 10,046.6 km^2(3,879 square miles) of open gulf water along its 998.4 km (624 miles) of coast. Unfortunately, monitoring water quality is a complex and difficult task, and Texas has not dedicated the needed resources to adequately monitor surface water quality or to monitor the health of the fish that roam these waters. A unique challenge has been assuring adequate water quality along the Texas-Mexico border, although pollution in water bodies occurs throughout the state.

All of these waters are afforded at least minimal amounts of protection by the state and federal governments by three different types of water quality standards:

1. stream standards, also referred to as surface water quality standards;
2. effluent standards (set for wastewaters); and
3. drinking water standards, which also cover groundwater used as a public water supply.

Along with Congress and federal water quality legislation like the Clean Water Act, the Texas legislature has recognized the need to protect water quality. In 1991 the legislature adopted a policy of "no net loss" of state-owned wetlands and authorized a state wetlands conservation management plan. Also in 1991 the legislature adopted the Clean Rivers Act, which directed the river authorities to conduct a regional assessment of water quality for each major river basin, with the Texas Natural Resource Conservation Commission overseeing the effort.

The Clean River Act supports the TNRCC's overall efforts to move water pollution management to a river basin or "watershed" approach. For example, the TNRCC has rewritten its rules to require that all permit renewals pertaining to a given river basin be considered in the same year. In this way, the TNRCC can take a river-basin-by-river-basin approach and better ensure that water quantity and quality are being maintained in the whole watershed.

Today, the Texas Natural Resource Conservation Commission is the primary agency responsible for water quality management in Texas, although it shares the responsibility with other state agencies such as the Texas Parks and Wildlife Department and the Railroad Commission of Texas.

Pollution has to some degree impacted all of Texas' 15 inland river basins and 8 coastal basins, several of its reservoirs, and all of its estuaries, coastal wetlands, and bays. According to the EPA's 1998 Water Quality Inventory—based upon TNRCC data—only 66 percent of the number of river miles with specific state standards fully supported the uses for which they were designated by the state.

In general, overall river and stream water quality remained remarkably similar between 1994 and 1998. Of the 6,864 km (4,290 miles) of rivers and streams that did not fully meet their designated use in 1996, 5,793 km (3,600 miles) did not meet safe swimming ("contact recreation") conditions, 1,921 km (1,194 miles) did not meet standards for aquatic life, and 19 km (12 miles) could not fully support boating and noncontact recreation uses.

Between 1994 and 1998, overall use support in reservoirs declined from 98 to 78 percent, indicating a substantial decline in reservoir water quality. The decline in overall use support was caused by higher levels of dissolved oxygen, higher levels of metals and organic substances, and elevated fecal coliform bacteria densities. Finally, the issuance of consumption advisories and aquatic life closures by the Texas Department of Health increased the number of reservoirs determined to yield fish that could not be safely consumed. Some 36,660 hm^2 (336,600 acres) of reservoirs were covered by fish-consumption advisories, while 500 hm^2 (500 acres) of reservoirs were also determined to yield fish unsafe for consumption and were subject to aquatic life closures.

While all reservoirs used as public water supplies did fully support this use, some reservoirs did not meet secondary drinking water standards for chloride, sulfate, and total dissolved solids, requiring expensive treatment processes to support this use. In all, 11 of the 99 classified reservoirs had elevated levels of one or more of these three secondary drinking water standards.

Vocabulary

perennial	终年的,长期的	assessment	评估
wetland	湿地,沼泽地	advisory	顾问的，咨询的，劝告的
water body	水体		
legislature	立法机构	coliform	大肠菌

Unit 4 Public Water Supply System

A public water supply system provides piped water for human consumption to 15 or more service connections or an average of at least 25 individuals each day for at least 60 days each year. The system includes the source water intake (such as a well), treatment, storage, and distribution piping. This definition of a public water supply system was specified by law as part of the federal Safe Drinking Water Act. Human consumption of water includes drinking water and water used for cooking, food preparation, hand washing, bathrooms and bathing. A private home served by its own well is not a public water supply system since it serves only a single service outlet. A mobile home park with 15 or more service connections is a public water supply system. A bar, restaurant or motel served by its own well is usually a public water supply system since it serves an average of 25 or more people each day, even though they may not be the same people every day. Schools and industries that have their own well are also public water supply systems if they have an average of 25 or more people at the facility each day. Seasonal establishments such as campgrounds and ski areas are also public water supply systems if they are open at least 60 days each year and serve an average of at least 25 people each day.

Purpose of Public Water Supply Systems

The main purpose of public water supply systems is to provide water which is safe for human consumption. Other important purposes are to provide an adequate quantity of water of acceptable taste, odour and appearance; and often to meet limited irrigation needs and fire protection. Providing water service places owners and operators of water systems under an ethical and legal obligation to meet these needs. Most people in the United States take safe, inexpensive drinking water for granted. We assume all water that comes from a tap is okay to drink, whether in a restroom, a gas station or a friend's home. Few of us realize the planning, monitoring, repair and maintenance required to obtain and protect adequate amounts of safe water. The federal government, through the Environmental Protection Agency (EPA), and individual states have established minimum requirements for water quality that must be met by public water supply systems. These requirements are meant to protect the public from contaminants which may cause acute or chronic health effects.

Contaminants that may have an immediate impact on health after drinking small amounts of water must be dealt with in all public water systems. These are contaminants that cause acute health effects. Examples are disease-causing organisms and nitrates. Contaminants that cause health effects if consumed over long periods of time must be dealt with in systems where the same residential or non-residential consumers have access to the water on a long-term basis. These are contaminants that cause chronic health effects. Examples include cancer-causing chemicals and chemicals affecting the nervous system or kidneys. The provision of an adequate quantity of water is addressed by properly sizing the source, pumping equipment, treatment, storage and piping to meet a reasonable demand for water created by all intended purposes. Taste, odour and colour are addressed through recommended maximum levels of certain contaminants that may make water unappealing.

Types of Public Water Supply Systems

Public water supply systems fit into three categories:

- Serves residences,
- Serves the same non-residents for most of each year, or
- Serves individuals who would only consume the water on a short-term basis.

The categories have been specified to relate the public health effects of potential contaminants to the risk of exposure to those contaminants. That is, if water from a particular system is consumed by the same individuals over a long period of time, then that system must ensure exposure to contaminants causing both acute and chronic health effects are addressed. If a system serves consumers only for a short period of time, contaminants causing acute health effects must be addressed. The three categories of public water supply systems are:

- Community,
- Non-transient non-community, and
- Transient non-community systems.

Community water systems serve residents. There must be 15 or more service connections on the same system or at least 25 residents served by the system. This category includes mobile home parks, subdivisions, water user associations, water districts, cities and towns, and some apartment buildings. Because people usually consume large amounts of water at their residences for many years, these systems must be concerned with contaminants having both acute and chronic health effects.

Non-transient non-community water systems serve schools and businesses which serve the same non-resident persons each day for more than six months per year. These systems must also be concerned with contaminants having both acute and chronic health effects.

Transient non-community water systems serve non-residents who do not work or attend school at the same facility for at least six months in a row. This category covers bars, restaurants, rest stops and campgrounds, to name a few. Transient non-community systems must be concerned with contaminants that may cause acute health effects by consuming very little water. They are not required to monitor and control contaminants that may cause long-term health effects since people are expected to consume the water for short periods of time.

Vocabulary

intake	(水管、煤气管等的)入口,进口
storage	储存
irrigation	灌溉
ethical	伦理的
acute	严重的
chronic	长期的,慢性的
residential	住宅的,居住的
cancer-causing	致癌的
have access to	可以利用,能利用
kidney	肾
long-term	长期的
subdivision	细分,再分,分部,小节

Reading Material A

Public Water Supply System Regulations

Regulations affecting public water supply systems cover a wide range of subjects. The following are the main areas which concern drinking water:

- Monitoring and reporting requirements deal with water quality, treatment, and public communication;
- Operator certification requirements pertain to individuals in responsible charge of a water system;
- Minimum design standards and the plan review and approval process ensure water system components are adequately sized and properly installed; and
- Other regulations address such issues as safety during system repair or construction, fire codes, and cross-connection control programs.

General information on each of these subjects will be presented in this chapter. Some items will be covered in more detail later in the manual. Specific monitoring and reporting requirements have intentionally been left out of this manual because they differ by system size and, in some cases, by specific system. Regulatory summaries and system-specific monitoring schedules should be obtained from the Public Water Supply Program, Montana Department of Environmental Quality.

Introduction to Montana Public Water Supply System Regulations

Montana has had a public water supply program since 1907, when outbreaks of waterborne disease and associated deaths moved the legislature to pass the first law regulating public water supplies.

Prior to 1970, protection of drinking water on a national level was the responsibility of the United States Public Health Service (PHS) which established standards for the quality of water used in interstate commerce.

In 1974, the federal government passed the Safe Drinking Water Act (SDWA). This act established national drinking water regulations to protect public health. The regulations addressed not only public health problems resulting from drinking water, but also health problems resulting from skin contact with the water and inhaling contaminant vapors released from the water.

Individual states, like Montana, are expected to carry out and enforce these regulations for public water supply systems. This role of implementation and enforcement by states is called primacy, for "primary enforcement authority". Montana must adopt regulations no less stringent than the federal requirements to maintain the role of primacy agency.

The Montana Department of Environmental Quality (DEQ) is responsible for implementing the Safe Drinking Water Act in Montana. The United States Environmental Protection Agency (EPA) oversees the state program to ensure minimum requirements are met. The federal Safe Drinking Water Act must be reauthorized by Congress every nine years. When Congress evaluates the adequacy of the Act during reauthorization, changes or amendments to the Act are often made.

The 1986 Amendments to the federal Safe Drinking Water Act significantly increased the number of contaminants public water supply systems must monitor. It tightened the requirements for systems which use surface water, and defined public notification requirements when monitoring is not performed or when a contaminant exceeds the maximum allowable limit.

The 1996 Amendments to the federal Safe Drinking Water Act added some flexibility for state implementation of the federal requirements but also included new items for certified operators and system capacity. "Capacity" as it is used here, refers to the financial, managerial, and technical abilities of a public water supply system. It is also sometimes referred to as system "viability".

Purpose of Public Water Supply System Regulations

Regulations governing public water supply systems serve two purposes. The primary purpose is to ensure reasonable protection of the health of people who consume the water (referred to as "protection of the public health"). A secondary benefit is that they help ensure protection of the investment dollars spent on construction of the public water supply system.

Public health protection is obtained by: 1) setting maximum contaminant levels (MCL's) for certain contaminants which may not be exceeded by a public water supply system; 2) ensuring monitoring for contaminants is done in a reasonable fashion; and 3) requiring treatment be installed to remove contaminants to below levels specified by their MCL.

The MCL for each contaminant is the enforceable drinking water standard, or primary standard. It is based on a maximum contaminant level goal (MCLG), a level below which no adverse health effects are expected to occur from drinking contaminated water. MCL's are set as close to the MCLG's as possible, taking costs and technology into consideration.

Some contaminants, for which analytical methods are poor or impractical, have minimum treatment requirements instead of MCL's. Examples of treatment technique requirements include filtration of surface water sources and corrosion control.

The "Multiple Barrier Concept" of public health protection incorporates several independent steps to provide public health protection. The theory behind this concept is the more barriers between a contaminant and the consumer, the more likely an isolated failure in one of the steps will not result in adverse public health effects.

For a public water supply system using ground water, steps in the multiple barrier concept include the following:

- Selection of the best source or source location;
- Development and implementation of a source water protection plan;
- Providing adequate treatment technology and eliminate contaminants;
- Monitoring water quality to check the effectiveness of treatment or the occurrence of contaminants (there are also often multiple barriers within treatment processes);
- Providing sanitary surveys to identify deficiencies which might impact water quality or service; and
- Reporting to the public any contamination events, monitoring failures, or water treatment deficiencies.

Proper design and construction of a public water supply system has a critical role in public health protection. It is also an expensive process regardless of the size of the system. Investment dollars are protected if the system is engineered, constructed, operated and managed so that it is able to provide safe water for as long as possible.

Monitoring water quality indicates if part of the system has failed, is leaking or is exposed to conditions which may shorten its useful life. Conditions which may affect the life of pipe include very hard water which might plug pipes, or corrosive water which "eats away" at the interior of pipes and tanks.

Vocabulary

certification	证明	amendment	修正,改正
pertain	适合,属于	shorten	缩短,(使)变短
standard	标准	multiple	多样的,多重的
outbreak	爆发	barrier	屏障,障碍物
waterborne	水生的	regardless of	不管, 不顾
inhale	吸入	interior	内部,内部的
primacy	首位		

Reading Material B

Ground Water Movement

Effect of Aquifer Characteristics

The ability of an aquifer to receive, store, or transmit water or contaminants depends on the characteristics of the aquifer. This includes the characteristics of the confining layers of a confined aquifer or the overlying unsaturated zone of an unconfined aquifer.

Hydraulic gradient, porosity and hydraulic conductivity are important concepts which determine ground water movement. Each of these are discussed briefly. More detail is available from references such as the Source Water Protection Technical Guidance Manual available from DEQ.

Ground water generally moves quite slowly—from about several feet per day to several feet per year-although it can move much faster in very permeable soils or in certain geologic formations, such as cavernous limestone. Gravity and pressure differences are also important factors in ground water movement. The direction and speed that ground water and accompanying contaminants flow are to a large degree determined by the hydraulic gradient.

The hydraulic gradient is the slope of a water table, or in a confined aquifer, the slope of the potentiometric surface (the surface defined by the elevation to which water rises in a well open to the atmosphere—also called the piezometric surface). In many cases in unconfined aquifers, the hydraulic gradient parallels the slope of the land surface.

Porosity refers to the amount of space between soil or rock particles and reflects the ability of a material to store water. Soils are said to be porous when the percentage of pore space they contain is large (such as a soil with porosity of 55 percent).

Hydraulic conductivity is a term that describes the ease with which water can pass through deposits and thus transmit water to a well. Generally, the larger the pores, the more permeable the material, and the more easily water can pass through.

Coarse, sandy soils are quite porous and permeable, and thus ground water generally moves through them rapidly. Bedrock is often not very porous, but may contain large fractures through which ground water passes quickly. Clay soils are quite porous but not very permeable and water moves through clay very slowly.

Fractures in consolidated rock play an important role in ground water movement. The fractures allow water to flow through them in many directions. This makes it difficult to predict and measure ground water flow in these formations.

Aquifers composed of limestone and other water-soluble rocks often have fractures which have been widened by physical or chemical erosion to form sinkholes, caves, tunnels or solution channels. Water and any accompanying contaminants often move very rapidly in these aquifers.

Effect of Well Pumping

Well pumping alters the natural movement of ground water. The depth from ground level to the top of the aquifer in a well not being pumped is called the static water level. When pumped, ground water around the well is pulled down and into the well. The depth from ground surface to the water level in the well during stabilized water withdrawal is called the pumping water level. The difference between the static water level and the pumping water level is called the drawdown. The greater the discharge of water from a well, the greater the drawdown experienced by the well.

The underground area affected by pumping is in the shape of an inverted cone and is called the cone of depression; the same area as viewed on a map of the ground surface is known as the zone of influence. The cone of depression may extend from a few feet to many miles, depending on local hydrogeological conditions. Locating additional wells within this zone of influence will cause the wells to compete with each other for water.

The zone of contribution is the area of the aquifer that recharges the well. The zone of contribution can also be altered by increased or decreased pumping. Any contaminants located in the zone of contribution might be drawn into the well along with the water.

Recharge of Aquifers

Replenishment of aquifers is known as recharge. Unconfined aquifers are recharged by precipitation percolating down from the land's surface. Confined aquifers are generally recharged where the aquifer materials are exposed at the land's surface—called an outcrop.

Surface waters also provide ground water recharge under certain conditions in many areas of Montana. When a surface water loses water to the adjacent aquifer, the stream is called a losing stream. When the opposite occurs and water flows from the ground water to the stream, it is called a gaining stream.

Properly identifying the recharge area of an aquifer is critical because the introduction of contaminants within the recharge area can cause aquifer contamination.

Knowing if the aquifer is influenced by a gaining or losing stream helps identify periods when biological contaminants from the surface water might reach the well water. Periods of gaining and losing stream flow may change seasonally, depending on the level of the ground water table. Monitoring the surface water level or stream stage and comparing it to the static water level in the well can give an indication of the direction of water flow.

Vocabulary

characteristic	特有的，典型的,特性，特征
unsaturated	不饱和的
unconfined	无约束的,无限制的
gradient	梯度
porosity	多孔的
permeable	有浸透性的，能透过的
geologic	地质(学)的,地质(学)上的
piezometric	压力计的,量压的,压力的
bedrock	岩床，根底，基础
fracture	破裂，(使)破碎
limestone	石灰石
drawdown	(抽水后)水位降低，水位降低量
replenishment	补充,补给
recharge	再充,回注

Unit 5 Assessment of the Drinking Water Supply

A comprehensive multi-barrier drinking water program includes:

- Source water protection;
- Sanitary surveys of the source area and distribution system to identify and prioritize risks to health;
- Watershed or well—head protection plans;
- Expansion capacity for forecasted population growth;
- For treated water, continuous optimal treatment;
- Routine maintenance of the distribution system;
- Treatment plant and distribution system classification, operator training and certification.

The assessment of the drinking water supply forms the basis of all activities related to providing the cleanest, safest, most reliable drinking water to the public. Assessments identify the characteristics of the water source, potential hazards, how these hazards create risks and how these risks can best be managed. The drinking water supply includes everything from the collection of the raw water to the point where the water reaches the consumer.

Selecting Source Waters

The first step in implementing a drinking water program is to identify water to be used as the source of drinking water. At this stage, an assessment should be made of the potential risks associated with the source. Risks could include wildlife in a watershed, recreational activities such as boating in a reservoir, wastewater treatment plants nearby, agricultural or industrial activities, etc. The system should be designed to minimize the impact of the risks over time. The characteristics of the source water—including physical (such as aquifer characteristics for groundwater sources), microbiological and chemical parameters, and the types of natural and anthropogenic contaminants present—determine the type of treatment required in order to deliver the cleanest, safest, most reliable drinking water to the public. Wherever possible, the system chosen should also be capable of being adapted to deal with unforeseen contaminants.

Groundwater Wells, Intakes and Raw Water Reservoirs

Some of the key characteristics in the design and the construction of groundwater wells, intakes and raw water reservoirs are the location, size and capacity, seasonal variations, retention times, design period, etc. In assessing these components, all potential hazards and their causes should be identified, and the level of risk associated with each of the hazards estimated, so priorities for risk management action can be established.

Treatment System

Treatment systems should be designed based on the site—specific raw water quality. Seasonal variations should be taken into account. Characteristics include treatment processes, treatment components, equipment design, chemicals used, treatment efficiency, monitoring procedures, etc. The treatment selected should address all potential hazards and the level of risk associated with those hazards.

Treated Water Reservoir, Service Connections and Distribution System

Treated water reservoirs and distribution systems should be designed to take the following into account: access by wildlife and people, system capacity, emergency water storage, contact time required for disinfection, minimizing or eliminating dead ends, and cross-connection controls. They should also be designed and constructed in compliance with all local or provincial by-laws, best management practices and regulations.

Vocabulary

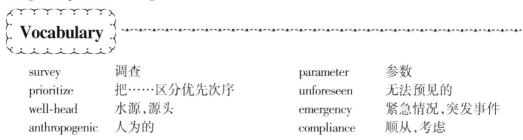

survey	调查	parameter	参数
prioritize	把……区分优先次序	unforeseen	无法预见的
well-head	水源,源头	emergency	紧急情况,突发事件
anthropogenic	人为的	compliance	顺从,考虑

Reading Material A

Drinking Water Quality Management

Regulatory Control

Regulatory agencies set, and ensure compliance with standards. Waterworks projects should be reviewed and, once approved, should have conditions to be met clearly outlined. Treatment plants should be inspected on an on-going basis to ensure quality benchmarks are being met. If these benchmarks are not being met, processes should be in place to remediate the situation. Compliance tools will vary from jurisdiction to jurisdiction.

Operational Procedures

The proper maintenance and operation of water supply, treatment and distribution systems are essential parts of any effort to ensure the on-going production and delivery of the highest quality drinking water possible. Operational procedures vary between treatment plants and between jurisdictions, but, generally speaking, operational-related monitoring requirements should be in place and clear; plants should be supervised by trained and certified operators; operator training programs should be available; facilities should be inspected on a regular basis; and administrative support should be available.

Monitoring, Reporting and Record-keeping

Monitoring drinking water quality occurs on a number of levels. Protocols should be in place for all activities, including selection of laboratories, routine monitoring, sample analysis and public notification. In general, monitoring requirements are specified by regulatory agencies.

Routine monitoring entails taking samples of raw water at the intake water at the treatment plant or from wells, and treated water in the distribution system at predetermined intervals to verify the quality of the water. Monitoring results should be reported directly to the drinking water authority as well as be available to the public. As mentioned previously, it is imperative for a reporting system to be in place for notifying the public when test results show drinking water presents a potentially serious health risk, or to explain the significance of changes in aesthetic quality. It is particularly important to have protocols in place dealing with the microbiological quality of the drinking water.

Other types of monitoring include on-going assessment of the location of sampling sites. Because samples are taken from such a small fraction of the water in any given system, as much as possible should be done to ensure the water in the samples is representative of the quality of the water throughout the plant and distribution system. In order to quickly remediate situations where water flow appears to be restricted, it is imperative that up-to-date drawings of the distribution system be kept in an accessible location.

Treatment plants may opt to use computer technology to help monitor water quality and operational variables (such as water pressure) on an automated basis. In addition, in order to facilitate the exchange of information about the water supply, jurisdictions may wish to set up databases which can be accessed by multiple users. More and more, members of the public are expecting to be able to access information over the internet about their water supply which may affect their health.

Certification and Training

Because treatment plant and distribution system operators have a significant degree of control over the quality of a community's drinking water, and thus over public health, appropriate and up-to-date training is essential. This training must include basic education about the need for disinfection to ensure public health goals are met.

Drinking water supplies in Canada and the United States are classified into categories based on size and complexity of operation. These classifications are used as the basis for training and certification programs for treatment plant operators.

Operator certification programs should be available to ensure treatment plant operators have appropriate levels of education, experience and knowledge to allow them to competently operate the type of plant they are working in.

It is imperative that operators and other staff have on-going access to opportunities for maintaining and upgrading their skills and knowledge on a regular basis.

Incident and Emergency Plans

Every system must have a set of procedures to follow in the event of incidents and emergencies. These procedures should be in place well in advance of any event. Plans should cover off any number of incidents, such as loss of source water, major main breaks, vandalism, power failure and deliberate chemical or biological contamination of the distribution system or reservoirs. Emergency plans should include clear procedures for the remediation of the situation and communication with appropriate authorities.

Evaluation and Audit

Any system as large and important as delivering clean, safe and reliable drinking water requires evaluation to ensure services are being delivered as planned and expected. For Canadian drinking water programs, evaluations verify that the elements detailed in this framework have been implemented properly and are being carried out effectively. The results of the evaluation are used as the basis for making improvements in future years.

Formal auditing can be carried out in the following areas:

- Distribution system audits;
- Construction audits;
- Operational audits;
- Regulatory compliance audits;
- Treatment performance audits;
- Administrative compliance audits;
- Water quality data audits;
- Design professional audits.

Public Awareness and Involvement

The public has expectations of government transparency, especially about issues that affect its health. As noted earlier, public involvement in the drinking water program is key to its success. Involving the public at every stage means:

- Making monitoring results or summaries available and easily accessible, such as on the internet;
- Notifying the public about risks to their health and what the authority is doing to address the risks;
- Issuing regular reports about drinking water systems, including improvements and areas which need further attention;
- Educating the public on a number of issues, including: the benefits of disinfection over the risks of microbiological contamination and disease; how guidelines are developed and what the values mean; how to prevent deterioration of water quality in the home; and the true cost of providing safe drinking water;

● Incorporating public consultations into decision-making processes which have an effect on public health, including the development process for new guidelines and regulations;

● Education about water conservation issues.

In the area of boil water advisories, members of the public must be informed when an advisory has been issued for their community, be given detailed information about the reason(s) for the advisory (whether it is precautionary or in response to an outbreak), and be told how long it is expected to be in place. Authorities should also consider visitors to their community when issuing an advisory—frequent advertising in highly visible areas may be prudent.

Private well owners need to be made aware that they are responsible for the quality of their own water, and that this should be tested regularly. This also applies to owners of private surface waters who use these as a source of drinking water. Owners need to know what to do should microbiological contamination occur or chemical contaminants be found in their drinking water, and how to properly abandon wells that are no longer safe or needed.

Vocabulary

benchmark	基准	jurisdiction	管辖(部门)
remediate	纠正	vandalism	故意破坏
procedure	程序,手续	transparency	透明,透明度
imperative	必要的	deterioration	变坏, 退化
representative	有代表性的	incorporate	合并
opt	选择	abandon	抛弃,放弃

Reading Material B

The Effects of Urbanization on Water Quality

You can understand why the water quality of our urban water supplies is so important. After all, the majority of the United State's population now live in or near cities. Big cities mean big development over large areas, which can certainly have an impact on the local water supply.

It's not hard to imagine that as cities grow, things happen that can harm the quality of the local water resources. That is why most governments must take measures to protect river, streams, lakes, and aquifers when small towns grow into big cities.

Here are some water-quality issues that relate to urban development.

Population Growth

If you live in a major city, you will see the effects of population growth every day. When more people move into an area a whole slew of support facilities must be built: housing developments, roads, shopping areas, and commercial and industrial facilities. Not only is land disturbed when development occurs, but the stress on the water resources of the region is increased to supply everyone with water.

The Atlanta, Ga. area is a good example of a booming urban center as it is one of the fastest growing metropolitan areas in the United States. Much of the growth and construction is occurring north of the city, which is where much of the water supply for Atlanta comes from. The population of the metro area has more than tripled from the 1 million residents in 1950 to over 3 million today, with no slowdown in sight. Ironically, development in the late 1980s was restricted because water-supply systems were not in place to handle the exploding growth in northern Atlanta. And for much of 1997, the city faced daily fines for releasing wastewater that had higher bacteria levels than were permitted.

Erosion and Sedimentation

The topography (lay of the land) of an area's watershed can have a lot to do with how water resources are affected by development. Large-scale development means that a lot of land clearing and grading occurs. If the area has sloping land, soils that erode easily, and receives frequent periods of heavy rainfall, then water quality can be affected, usually in a negative way.

Eroded soil from construction sites is carried to streams and lakes where it causes 1) excess turbidity that harms aquatic life, increases water-treatment costs, and makes the water less useful for recreation; and 2) sedimentation that clogs drainage ditches, stream channels, water intakes, and reservoirs, and destroys aquatic habitats.

Erosion and sedimentation controls include a wide range of temporary and permanent measures. Planting vegetation is one of the best ways to stabilize soils and minimize erosion. Sedimentation ponds are used to capture storm runoff and allow sediment to settle to the pond's bottom, and silt fences are used to contain water runoff and minimize sedimentation in nearby streams. As the picture to the right shows, silt fences do not always work.

Without proper design, installation, and maintenance of erosion and sedimentation controls, sediment-laden runoff from construction sites can damage streams and nearby properties. In this case, a tributary draining a construction site is depositing sediment-laden water into the clearer river at the top.

Urban Runoff

Much of the rainfall in watersheds having forests and pastures is absorbed into the porous soils (infiltration), is stored as ground water, and moves back into streams through seeps and springs. Thus, in many rural areas, much of the rainfall does not enter streams all at once, which helps prevent flooding.

When areas are urbanized, much of the vegetation and top soil is replaced by impervious surfaces such as roads, parking lots, and pavement. When natural land is altered, rainfall that used to be absorbed into the ground now must be collected by storm sewers that send the water runoff into local streams. These streams were not "designed by nature" to handle large amounts of runoff, and, thus, they can flood. Drainage ditches to carry stormwater runoff to storage ponds are often built to hold runoff and collect excess sediment in order to keep it out of streams.

So, how can excessive urban runoff harm streams?

1. Water running off of impervious areas, such as roads and parking lots, can contain a lot of contaminants, such as oil and garbage. This runoff often goes directly into streams.

2. Following summer storms, runoff from heated roads and parking lots causes rapid increases in stream temperatures that can produce thermal shock and death in many fish.

3. Use of stormwater impoundments and porous paving materials can reduce stormwater runoff and the movement of contaminants from roads and other areas to streams.

4. Regulations and controls on the location and amount of impervious area can lessen the damage that contaminants can do to streams.

5. Runoff of sand and salt used to help remove snow from roads can contaminate streams.

Nitrogen

Nitrogen, in the forms of nitrate, nitrite, or ammonium, is a nutrient needed for plant growth. About 78 percent of the air that we breathe is composed of nitrogen gas, and in some areas of the United States, particularly the northeast, certain forms of nitrogen are commonly deposited in acid rain. Although nitrogen is abundant naturally in the environment, it is also introduced through sewage and fertilizers. Chemical fertilizers or animal manure is commonly applied to crops to add nutrients. It may be difficult or expensive to retain on site all nitrogen brought on to farms for feed or fertilizer and generated by animal manure. Unless specialized structures have been built on the farms, heavy rains can generate runoff containing these materials into nearby streams and lakes. Wastewater-treatment facilities that do not specifically remove nitrogen can also lead to excess levels of nitrogen in surface or ground water. Two of the major problems with excess levels of nitrogen in the environment are:

1. Excess nitrogen can cause over-stimulation of growth of aquatic plants and algae. Excessive growth of these organisms, in turn, can clog water intakes, use up dissolved oxygen as they decompose, and block light to deeper waters. This seriously affects the respiration of fish and aquatic invertebrates, leads to a decrease in animal and plant diversity, and affects our use of the water for fishing, swimming, and boating.

2. Too much nitrate in drinking water can be harmful to young infants or young livestock.

Phosphorus

Phosphorus is an essential element for plant life, but when there is too much of it in water, it can speed up eutrophication (a reduction in dissolved oxygen in water bodies caused by an increase of mineral and organic nutrients) of rivers and lakes. This has been a very serious problem in the Atlanta, Ga. area, as a major lake that receives Atlanta's wastewater, West Point Lake, is south of the city. In metropolitan Atlanta, phosphorus coming into streams from point source, primarily wastewater-treatment facilities, have caused West Point Lake to become highly eutrophic ("enriched"). A sign of this is excess algae in the lake. State laws to reduce phosphorus coming from wastewater-treatment facilities and to restrict the use of phosphorus detergents has caused large reductions in the amounts of phosphorus in the Chattahoochee River south of Atlanta and in West Point Lake.

1. Towns in the metropolitan Atlanta area are continuing to expand and upgrade existing wastewater-treatment facilities to handle the increasing volume of wastewater and sewage and to meet stiffer regulations on effluent and river quality.

2. Additional control of phosphorus from non-point sources (such as applications of lawn fertilizers and disposal of animal wastes) may be useful to maintain or improve the water quality in streams and lakes near growing urban areas.

Sewage Overflows

Many sewer lines are constructed next to streams to take advantage of the continuous, gradual slopes of stream valleys. Blockages, inadequate carrying capacity, leaking pipes, and power outages at pumping stations often lead to sewage overflows into nearby streams. There are three types of sewer systems:

1. Sanitary sewers carry storm runoff from streets, parking lots, and roofs through pipes and ditches, and eventually into streams.

2. Sanitary sewers carry raw sewage from homes and businesses to wastewater-treatment facilities.

3. Combined sewers carry a combination of raw sewage and stormwater runoff.

This picture of a sanitary sewage overflow illustrates a common problem concerning sewage overflows that occur in urban areas. Sanitary sewer overflows occur when sewer pipes clog or pumping stations break down. As shown here, raw sewage overflows from manholes and leaking pipes into nearby streams rather than backing up into homes and businesses.

Combined sewer overflows occur during storms when there is more stormwater flowing than the pipes leading to a treatment plant can handle. The excess runoff flushes human and industrial wastes, oil, toxic metals, pesticides, and litter into streams.

Waterborne Pathogens

Waterborne pathogens are disease-causing bacteria, viruses, and protozoans that are transmitted to people when they consume untreated or inadequately treated water. Two protozoans in the news today are Giardia and Cryptosporidium. Their consumption can lead to severe problems of the digestive system, which can be life-threatening to the very young, very old, or those with damaged immune systems.

Many cities routinely monitor urban streams to measure the amounts of bacteria that, although harmless themselves, have similar sources (animal and human waste) as do the waterborne pathogens. The harmless bacteria therefore act as indicators of the possible presence of other bacteria that are not harmless. Treated water coming out of wastewater treatment plants is also monitored for bacteria. And in some larger cities additional monitoring of drinking water has begun.

Toxic Metals

Small amounts of some toxic metals tend to accumulate in the food chain and can damage living things. In the past, most toxic metal pollution came from mining activities and individual sources, such as wastewater-treatment plants and smoke-stack emissions. Federal and State regulations have resulted in the reduction of toxic metals from these sources. Metals tend to attach themselves to dirt and sediment, and thus, they are still present in the stream beds and banks of many urban streams.

Concentrations of toxic metals in stream sediments could be reduced if

1. trash is properly disposed of or recycled;
2. major streets and parking lots are routinely cleaned;
3. stormwater is caught and kept in ponds to trap metal-laden sediments.

Pesticides

Pesticides are chemical and biological substances intended to control pests, such as insects, weeds, bacteria, and algae. Pesticides are heavily used on farmland, but in urban areas, the main usage is on residential and commercial properties. When storms hit, the runoff from yards and roadsides carry pesticides into local streams, where they may harm aquatic life and enter drinking-water supply intakes. In urban areas, the wise usage of pesticides is the key to reducing pesticide problems that are increasingly occurring in our drinking water. Pesticides should only be applied when necessary and as recommended by the product labels. Persons applying pesticides should avoid spreading the product onto pavements, gutters, curbs, and storm drains.

Vocabulary

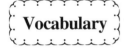

urbanization	都市化	nitrite	亚硝酸盐
booming	急速发展的	ammonium	铵
turbidity	浊度	use up	用完，耗尽
sedimentation	沉淀	decompose	分解，腐烂
ditch	沟渠	respiration	呼吸，呼吸作用
silt	淤泥	phosphorus	磷
impervious	不透水的	eutrophication	富营养作用
thermal	热的，热量的	sewage	污水，下水道
nitrogen	氮	protozoan	原生动物
nitrate	硝酸盐	indicator	指示剂

Unit 6 Pumps and Pumping Stations

Pumps and pumping machinery serve the following purposes in water systems: 1) lifting water from its source (surface or ground), either immediately to the community through high-lift installations, or by low lift to purification works; 2) boosting water from low-service to a high-service areas, to separate fire supplies, and to the upper floors of many storied buildings; and 3) transporting water through treatment works, backwashing filters, draining component settling tanks, and other treatment units, withdrawing deposited solids and supplying water (especially pressure water) to operating equipment.

Today most water and wastewater pumping is done by either centrifugal pumps or propeller pumps. How the water is directed through the impeller determines the type of pump. There is 1) radial flow in open-or closed-impeller pumps, with volute or turbine casings, and single or double suction through the eye of the impeller, 2) axial flow in propeller pumps, and 3) diagonal flow in mixed-flow, open-impeller pumps. Propeller pumps are not centrifugal pumps. Both can be referred to as rotodynamic pumps.

Open-impeller pumps are less efficient than closed-impeller pumps, but they can pass relatively large debris without being clogged. Accordingly, they are useful in pumping wastewaters and sludges. Single-stage pumps have but one impeller, and multistage pumps have two or more, each feeding into the next higher stage. Multistage turbine well pumps may have their motors submerged, or they may be driven by a shaft from the prime mover situated on the floor of the pumping station.

In addition to centrifugal and propeller pumps, water and wastewater systems may include 1) displacement pumps, ranging in size from hand-operated pitcher pumps to the huge pumping engines of the last century built as steam-driven units; 2) rotary pumps equipped with two or more rotors (varying in shape from meshing lobes to gears and often used as small fire pumps); 3) hydraulic rams utilizing the impulse of large masses of low pressure water to drive much smaller mass of water (one half to one sixth of the driving water) through the delivery pipe to higher elevations, in synchronism with the pressure waves and sequences induced by water hammer; 4) jet pumps or jet ejectors, used in wells and dewatering operations, introducing a high-speed jet of air or water through a nozzle into a constricted section of pipe; 5) air lifts in which air bubbles, released from upward directed air pipe, lift water from a well or sump through an eductor pipe; and 6) displacement ejectors housed in a pressure vessel in which water (especially wastewater) accumulates and from which it is displaced through an eductor pipe when a float-operated valve is tripped by rising water and admits compressed air to the vessel.

Pumping units are chosen in accordance with system heads and pump characteristics. The system head is the sum of the static dynamic heads against the pump. As such, it varies with required flows and with changes in storage and suction levels. When a distribution system lies between pump and distribution reservoir, the system head responds also to fluctuations in demand. Pump characteristics depend on pump size, speed, and design. For a given speed N in revolutions per minute, they are determined by the relationships between the rate of discharge, Q, usually in gallons per minute, and the head H in feet, the efficiency E in percent, and the power input P in horsepower. For purpose of comparison, pumps of given geometrical design are characterized also by their specific speed N_s, the hypothetical speed of a homologous (geometrically similar) pump with an impeller diameter D such that it will discharge 1 gallon/min (1 gallon = 3.785 L) against a 1-ft (1 foot = 0.3 m) head. Because discharge varies as the product of area and velocity, and velocity varies as $H^{1/2}$, Q varies as $D^2 H^{1/2}$. But velocity varies also as πDN/60. Hence $H^{1/2}$ varies as DN, or N varies as $H^{3/4}Q^{1/2}$.

Generally speaking, pump efficiencies increase with pump size and capacity. Below specific speeds of 1 000 units, efficiencies drop off rapidly. Radial-flow pumps perform well between specific speeds of 1 000 and 3 500 units; mixed-flow pumps in the range of 3 500 to 7 500 units; and axial-flow pumps after that up to 12 000 units. For a given N, high-capacity, low-head pumps have the highest specific speeds. For double-suction pumps, the specific speed is computed for half the capacity. For multistage pumps, the head is distributed between the stages. This keeps the specific speeds high and with it, also, the efficiency.

Specific speed is an important criterion, too, of safety against cavitation, a phenomenon accompanied by vibration, noise, and rapid destruction of pump impellers. Cavitation occurs when enough potential energy is converted to kinetic energy to reduce the absolute pressure at the impeller surface below the vapor pressure of water at the ambient temperature. Water then vaporizes and forms pockets of vapor that collapse suddenly as they are swept into regions of high pressure. Cavitation occurs when inlet pressure are too low or pump capacity or speed of rotation is increased without a compensating rise in inlet pressure. Lowering a pump in relation to its water source, therefore, reduces cavitation.

Vocabulary

pumping station	泵站	meshing	咬合
centrifugal	离心的	lobe	突齿,凸起
propeller	螺旋桨	ejector	喷射
propeller pump	螺旋泵、轴流泵	specific speed	比速
impeller	叶轮	eductor	喷射器
radial	径向的,轴向的	nozzle	喷嘴
volute	螺旋形的	cavitation	气蚀作用
casing	壳体,外壳	vibration	振动
diagonal	对角线的	kinetic	动力学的
rotodynamic	旋转动力的	ambient	周围的,环境的
shaft	轴	homologous	相应的,相似的
mover	马达,发动机		

Reading Material A

Pumps

A water system needs to move the water produced from the source to its customers, the end user. In almost all cases, the source is at a lower elevation than the user so the water must be raised to a higher level. Some type of pumping equipment must be used to generate the pressure for raising the water to the higher elevation.

Many different types of pumps can be used with the selection depending on the work that needs to be done. One type would be used for transferring water from a well to a tower; another would be better suited for pumping sludge containing a lime byproduct from a softening plant; still another would be used for feeding a chemical into the water for treatment. Among the considerations in selecting a pump are the maximum flow needed in gallons per minute (gpm), the head it needs to pump against, and the accuracy needed for flow control.

Positive Displacement Pump

The positive displacement pump is commonly used to feed chemicals into the water or to move heavy suspension, such as sludge.

Piston Pump

One type of positive displacement pump consists of a piston that moves in a back and forth motion within a cylinder. It is used primarily to move material that has a large amounts of suspended material, such as sludge, in wastewater. The cylinder will have check valves that operate opposite to each other, depending on the motion of the piston. One check will be located on the suction side of the piston and will open as the piston moves back, creating a larger cylinder area. After the piston has reached the longest stroke position, the motion of the piston will reverse. This action will open the discharge check valve and close the suction check. The contents of the piston are then discharged to discharge piping. After the discharge, the motion of the piston will reverse and the suction stroke will begin. This action will take place as long as power is applied to the pump.

Because high pressure could damage parts of the pump or cause the piping material to fail, the positive displacement pump should never be throttled on the discharge side of the piston. The power for a large piston pump is generally an electric motor connected to the piston by way of a gear head and connecting rod.

Diaphragm Pump

Another type of positive displacement pump used in the water industry is the diaphragm pump. This pump operates the same way as the piston pump except that, in place of a piston that moves in a cylinder, a flexible diaphragm moves back and forth in a closed area. The check valves operate in the same fashion as they feed or move liquid in the pump. This type of pump is used when high accuracy is required. Most of these pumps are operated by the use of a solenoid that will pulse a set number of times per minute. This pulsing, which is termed frequency, is variable and can be set by the operator or by a control signal. The length of the stoke can also be adjusted in order to vary the size of chamber that fills with liquid.

Centrifugal Pumps

Because it delivers a constant flow of water at a constant pressure for any given set of conditions, the centrifugal pump is ideal for delivering water to customers. Most well pumps are centrifugal pumps. They are ideal for use in the distribution system since they do not produce pulsating surges of flow and pressure.

This pump operates on the theory of centrifugal force. As the impeller rotates in the pump case, it tends to push water away from the center of the rotation. As the water is pushed away from the center of the impeller, additional water is pulled into the eye, or center, of the impeller. The water that has been pushed to the outside of the impeller is removed from the pump through the discharge piping. This water will have a pressure that is determined by the pitch of the impeller and the speed at which the impeller is turning.

Types of Centrifugal Pumps

There are four major types of centrifugal pumps. These include turbine pumps, low lift centrifugal pumps, high lift centrifugal pumps, and booster pumps.

A line shaft turbine pump is one of the most common pumps to be used in the water industry. The pump consists of bowls that contain the impellers, which are connected with each other through the pump shaft bearings. A turbine pump will usually be staged with more than one impeller to overcome the head conditions that are encountered in the operation. The water from one stage will be discharged into the suction eye of the next stage, a process that will continue until the head is overcome. The size or diameter of the first impeller dictates the volume capacity of the pump in gpm.

Low lift centrifugal pumps are used to pump water from at surface source to a treatment plant. High lift centrifugal pumps are used to pump water from a treatment plant to the distribution system. Booster pumps are used to increase water pressure in the water distribution system.

Centrifugal pumps can have more than one stage, or impeller. Each additional stage increases the head that the pump can pump against. If one impeller will pump against 18 m of head, two will pump against approximately 36 m, three against 55 m, etc. The rate of the flow in gpm will not be affected by additional impellers since that is dictated by the diameter of the impeller. It will be no greater than what the first impeller can deliver.

Motor Maintenance

Information on the maintenance of the motor should be available in the manufacturer manual supplied with the motor. It includes the types of oils or greases to be used and how often the equipment should be lubricated.

Maintenance of the insulation inside of the motor is a difficult, but important, task. It needs to be kept dry, cool, and clean, free of contaminants such as dust, salts, chemicals, lint, and oil. It is important to clean the vents to keep them open.

Pump Maintenance

The only maintenance needed for a line shaft turbine pump is checking the packing gland for excessive leakage and repacking as needed. If the leakage at the packing gland is excessive, the operator should tighten the packing gland follower until the water loss is reduced, but the leakage should not be completely stopped since water serves as the coolant for the packing in the stuffing box.

When the follower cannot be tightened anymore, the packing has to be removed and new packing installed. New packing should never be added on top of the old. It is now common to have a mechanical seal instead of packing. They are more expensive, but their maintenance is low.

Vocabulary

elevation	高程	diaphragm pump	隔膜泵
flow	流量	cylinder	圆柱体,圆筒
gpm	gallons per minute	suction	吸力,吸入
	(1 gallon = 3.785 L)	turbine pump	涡轮泵
head	压力	booster pump	增压泵
piston pump	活塞泵	insulation	绝缘

Reading Material B

Water-supply Engineering

A water supply for a town usually includes a storage reservoir at the source of the supply, a pipeline from the storage reservoir to the distribution reservoir near the town, and finally the distribution pipes buried in the streets, taking the water to the houses, shops, factories and offices. The main equipment is thus the two reservoirs and the pipeline between them. The function of the storage reservoir is keep enough water over one or several years to provide for all high demands in dry periods, and the distribution reservoir has the same function for the day or the week, the storage reservoir by its existence allows the supply sources to be smaller and less expensive. And the distribution reservoir similarly allows the pipeline and pumps to be smaller and cheaper than they would be if it did not exist.

In the United States, some of whose cities have the largest water use in the world per person, the average use per person varies from 200 to 5 000 litres per day, averaging some 500 litres per day per person. But it must not be assumed that colder countries will eventually reach the same level of use, because much of the highest US demand comes from the water spent in summer on air conditioning equipment and the watering of gardens.

Water engineers must therefore study the water use per person (consumption per head) in their own country and choose a figure based on the most advanced community there. The chosen consumption per head must be multiplied by the estimated population at the date for which the supply is being planned, some thirty years ahead or more. The supply and storage equipment must be designed to be large enough for this period since neither of them is so easily extended as the distribution system. This can be extended as the need arises and as the houses are built.

Once the volume of the required yearly supply has been calculated and agreed with all concerned, including the fire department, it is important to make sure that it really is permanently obtainable from the catchment area proposed. The catchment area is the area which drains towards the supply, and the yearly amount of water drawn off to the storage reservoir cannot be more than the rainfall on the catchment area and should usually be very much smaller.

A water supply may be obtained from surface water (rain) or from underground water or both. Both are refilled by the rainfall, the surface water by the runoff, and the springs or wells by the water which enters the ground, the infiltration water. These two quantities, plus the evaporation water and the water used by the trees and plants, make up the total rainfall. Even if the community water supply includes all the springs as well as the surface water in the area, it still does not obtain all the rainfall because of evaporation and the needs of plant life.

It is therefore important to check the rainfall records and the runoff and infiltration vales with the records of the stream flows and other local water information. Infiltration water, the rainfall which enters the ground and becomes ground water, can travel for long horizontal distances, and it may pass into or out of the catchment area. If the yearly water supply exceeds the yearly rainfall, the ground water level will generally fall and eventually it will become impossible to obtain the required supply. Another source will have to be found.

It is not essential to build a storage or impounding reservoir if the water can be stored in the ground, and this may often be possible. In face in the London area, where the water level in the chalk has been steadily falling for more than a century because of pumping, it has been suggested that further storage shall be not by the surface reservoirs which have been used until now, but by recharging the chalk with purified water through wells changed for the purpose.

This practice of underground storage is being widely used by the gas industry in many countries. Gas is sent underground by compressors through wells into a sealed underground sand, limestone, or other porous formation at a time when the gas supply is large, to be stored until the demand is larger than the supply. These underground containers for gas are often hundreds of times as large as the largest gas tank in existence and have been found to be a cheap, practical, and safe way of storing gas.

Vocabulary

distribution pipe	配水管道	catchment area	集水区
high demands	高峰用水量	impounding basin	蓄水池
to be multiplied by	乘以	purify	使纯净,使洁净
fire department	消防部门	porous	多孔的

Unit 7 Water Supply System

The natural of the water source commonly determines the planning, design, and operation of the collection, purification, transmission, and distribution works. The two major sources used to supply community and industrial needs are referred to as surface water and groundwater. Streams, lakes, and rivers are the surface water sources. Groundwater sources are those pumps from wells.

Fig. 7.1 depicts an extension of the water resource system to serve a small community. The source in each case determines the type of collection works and the type of treatment works. The pipe network in the city is called the distribution system. The pipes themselves are often referred to as water mains. Water in the mains generally is kept at a pressure between 200 and 860 kilopascals (kPa). Excess water produced by the treatment plant during periods of low demand (usually the nighttime hours) is held in a storage reservoir. The storage reservoir may be elevated (the ubiquitous water tower), or it may be at ground level. The stored water is used to meet high demand during the day. Storage compensates for changes in demand and allows a small treatment plant to be built. Storage is also used to provide emergency backup in case of a fire.

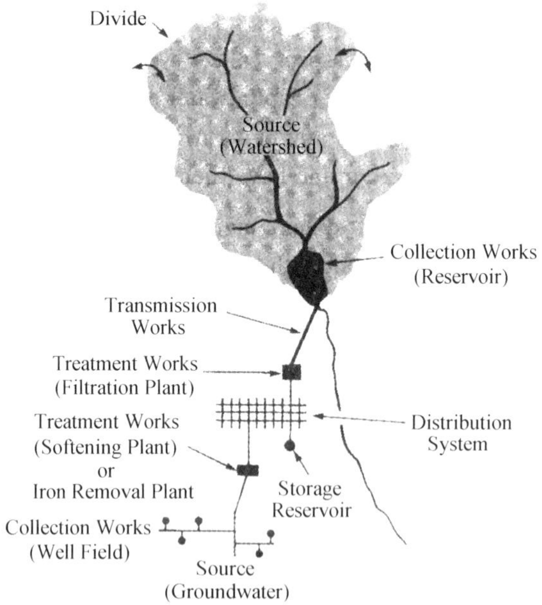

Fig. 7.1 An extension of the water supply resource system

Population and water consumption patterns are the prime factors that govern the quantity of water required and hence the source and the whole composition of the water resource system. One of the first steps in the selection of a suitable water supply source is determining the demand that will be placed on it. The essential elements of water demand include the average daily water consumption and the peak rate of demand. The average daily water consumption must be estimated for two reasons: 1) to determine the ability of the water source to meet continuing demands over critical periods when surface flows are low or groundwater tables are at minimum elevations; and 2) for purposes of estimating quantities of stored water that would satisfy demands during these critical periods. The peak demand rates must be estimated in order to determine plumbing and pipe sizing, pressure losses, and storage requirements necessary to supply sufficient water during periods of peak water demand.

Many factors influence water use for a given system. For example, the mere fact that water under pressure is available stimulated its use, often excessively, for watering lawns and gardens, for washing automobiles, for operating air-conditioning equipment, and for performing many other activities at home and in industry. The following factors have been found to influence water consumption in a major way.

1. Industrial activity.
2. Meterage.
3. System management.
4. Standard of living.
5. Climate.

The following factors also influence water consumption to a lesser degree: extent of sewerage, system pressure, water price, and availability of private wells. The influence of industry is to increase per capita water demand. Small rural and suburban communities will use less water per person than industrialized communities. Industry is probably the largest single factor influence per capita water use.

The second most important factor in water use is whether individual consumers have water meters. Meterage imposes a sense of responsibility in unmetered residences and business. This sense of responsibility reduces per capita water consumption because customers repair leaks and make more conservative water-use decision almost regardless of price. Because water is so cheap, price is not much of a factor.

Following meterage closely is the aspect called system management. If the water distribution system is well managed, per capita water consumption is less than if it is not well managed. Well-managed, are those in which the managers know when and where leaks in the water mains occur and have them repaired promptly.

Industry activity, meterage, and system management are more significant factors controlling water consumption than are either the standard of living or the climate. The rationale for the latter two factors is straightforward. Per capita water use increases with an increased standard of living. Highly developed countries use much more water than the less developed nations. Likewise, higher socioeconomic status implies greater per capita water use than lower socioeconomic status. Higher average annual temperature implies higher per capita water use, whereas areas of high rainfall experience lower water use.

The average national value for water consumption in the United States in 1970 was 628 liters per capita per day (Lpcd) A similar study conducted in 1954 yielded a value of 556 Lpcd. The average single-family residence uses about 208 L/d. The variation in demand is normally reported as a factor of the average day. For metered dwellings the factors are as follows:

$$\text{Maximum day} = 2.2 \times \text{average day}$$

$$\text{Peak hour} = 5.3 \times \text{average day}$$

But for large cities, the variation factors are probably between 1.2 ~ 1.7.

Vocabulary

purification	净化	kilopascal	千帕
pipe network	管网,管道系统	backup	备份,储备
distribution system	配水系统	meterage	(用仪表)计量
main pipe	干管	water meter	水表

Reading Material A

Distribution Systems

The distribution system consists of booster pumps, pipes, meters, storage tanks, control valves, and hydrants. This part of the system is often neglected because much of it is underground and out of sight, but it needs and deserves almost as much operator attention as source and treatment facilities.

An adequate distribution system is able to provide a sufficient amount of safe water to all users at a pressure that will satisfy normal needs. It also provides water without undue water loss.

Even in a metered system, not all water coming from the well can be accounted for. Some may be lost because of leaks in the system, some may be discharged through hydrants as part of a flushing program, some may evaporate from storage tanks or be used fighting fires. This portion is usually referred to as unaccounted-for water.

The total amount of water delivered to the customers is usually measured in million gallons per day (mgd), or in gallons per capita day (gpcd) which is the total number of gallons divided by the number of persons being served. An operator of a small water system may not be involved in the design, construction or repair of the distribution system. However, to properly maintain the system it is important that operators are familiar with all parts of the system and the effects they may have on the quality of the water served to customers.

Piping Materials

A variety of pipe materials are available for use in a public water supply system to carry water under pressure. They include plastic, ductile iron, steel and concrete. Service pipes—those running from the distribution system to a customer—may be made of copper, plastic, iron, steel, or brass.

All components used in a water distribution system must conform to the latest standards issued by the American Water Works Association (AWWA), or by the ANSI/NSF standards for potable water when AWWA standards don't exist (e.g., for less than 0.1 m diameter pipe). These organizations have established standards for water supply materials which ensure the materials used will not leach contaminants into the drinking water and meet minimum strength criteria. Used materials can be reused if they meet the standards and have been restored to practically original condition. The amount of water a pipe can carry is a function of its size and the smoothness of the interior surface.

Valves

Valves are a very important part of the distribution system because they regulate the flow of water, reduce pressure, provide air and vacuum relief, blow off or drain water from parts of the system and prevent backflow.

The operator should know the locations of these valves so they can be used when necessary. Each water system should have a map available with the location of the valves clearly marked. It is important that the access to the valves not be compromised by other facilities. For example, access to some valves have unknowingly been covered during a street repair process.

Most valves will need to be exercised at least on an annual basis. If not, they may become stuck and will be inoperable when really needed. This valve exercising process should be geared into a routine maintenance plan.

Valves of any size need to be operated slowly. Water hammer is caused by closing the valve too quickly. When water is suddenly stopped, shock waves are generated, which cause large pressure increases throughout the system. These shock waves travel quickly and can cause extensive damage, sometimes splitting pipes or blowing fittings completely off the system. Frequently in the operation of valves, conditions cause a partial vacuum or void to occur on the downstream side of a valve. These voids will fill with low-pressure vapors from the water. When these pockets implode or collapse, they create a mechanical shock causing pockets of metal to break away from the valve surface. A noisy or vibrating valve may be an indication that cavitation is occurring, which will eventually result in leaks and a valve unsuitable for service.

Pressure Reducing Valves

Pressure reducing valves reduce the water pressure by restricting the flow. Pressure on the downstream side of the valve regulates the amount of flow permitted through it. This type of valve is usually of the globe design with a spring-loaded diaphragm which sets the amount of the opening. As downstream pressure is exerted against the diaphragm, the spring is compressed, moving the valve element toward the seal, thereby limiting the flow. If the downstream pressure drops, the spring will open the valve element and provide more flow.

These valves allow distribution systems to maintain pressures to the customers in the desirable

344.5 to 551.2 kPa range. Pressure regulators are especially valuable in the mountainous portions of Montana where elevation differences produce great differences in pressure within the distribution system.

Fire Hydrants

If the water system is designed to provide fire protection, the water lines and the hydrant connections to the distribution system must be a minimum of six inches in diameter. The location and spacing of hydrants is specified in State design standards. Within a residential environment they should be at every street intersection. Flushing hydrants may be smaller than 6-inch diameter, but must not have fire-fighting compatible connections.

Storage Reservoirs

Small water systems may have a small amount of storage in the form of a hydropneumatic tank—either a standard pressure tank with an air/water interface, or a captive air tank. Other systems might have ground level concrete or steel storage tanks or elevated steel tanks which serve the system by gravity.

Small hydropneumatic pressure tanks, are used to maintain pressure within the distribution system and to prevent well pumps from cycling too frequently. Larger hydropneumatic tanks use on-site, permanent, air charging devices to maintain the needed air pressure within the vessel. The tank should have a sight glass so the air/water ratio in the tank can be observed and adjustments made.

Concrete or steel storage tanks will provide a greater amount of storage capacity than a hydropneumatic tank and may provide sufficient fire flows. They also will continue to operate for a period of time during a power outage.

Looped Systems and Dead End Mains

Water lines should be installed to loop back into another part of the distribution system. This allows circulation of water to all users. Dead ends should be avoided. The lack of movement of the water in these lines will cause stagnation and result in the growth of slimes and bacteria, and development of taste and odor problems.

If it is necessary to have a dead end line, a fire hydrant or a flushing hydrant should be installed on the end of the line so stagnant water can be routinely discharged. The hydrant should be large enough to generate a flow of 0.76 m per second in the line for thorough flushing.

Leak Detection Program

Leak detection programs are an effective way to reduce operating and maintenance costs. If leaks can be detected when they are small, the system may save many dollars, hours of work and possible property damage. Leaks not only waste water, but may create an environment around the pipe which increases corrosion. Once corrosion develops pinhole leaks in the pipe, contamination can be drawn into the system when the pressure in the pipe is reduced.

If the water system is completely metered, a water balance between the amount produced and the amount billed to customers can sometimes indicate when losses are occurring.

Vocabulary

booster pumps	增压泵	cavitation	气穴现象
hydrant	消防栓,消防龙头	hydropneumatic	液压气动的,水和空气协同作用的
undue	不适当的		
latest	最近的	stagnation	停滞
water hammer	水锤		

Reading Material B

Dual Water Distribution

As the name implies, dual distribution systems involve the use of water supplies from two different sources in two separate distribution networks. The two systems work independently of each other within the same service area. Dual distribution systems are usually used to supply potable water through one distribution network and non-potable water through the other. The systems would be used to augment public water supplies by providing untreated, or poorly treated, water for purposes other than drinking. Such purposes could include fire-fighting, sanitary flushing, street cleaning, or irrigation of ornamental gardens or lawns. This system has been used in U.S. Islands.

Technical Description

The systems are designed as two separate pipe networks: a potable water distribution system, and a system capable of distributing sea water or other non-potable waters. The system includes distribution pipes, valves, hydrants, standpipes, and a pumping system, if required. Pipes in the systems are generally cast iron or ductile iron, although fiberglass has also been used.

In seawater-supplied systems, pumps are required to lift the seawater to higher elevation storage tanks. Likewise, pumps may be required to lift wastewaters from wastewater sumps or other collection points. The pumping systems consist of a pumping station containing the water intake, a pumping well, and an elevated storage tank for emergency use. The pumps require foot valves, or one-way valves, in order to retain their charge of water. The water is pumped through a manifold into the secondary or alternative distribution system.

The potable-water, or primary, system operates like any other potable-water supply and distribution system, requiring a water source, treatment plant, storage facility, and distribution system. Pumps are generally required to lift potable water from the treatment plant to storage tanks, from which it is distributed by gravity to the point of use.

This technology is rarely used. Seawater-based systems have been used in Castries, Saint Lucia, for fire-fighting purposes and in Charlotte Amalie, U. S. Virgin Islands. U. S. Navy bases have installed and operated similar systems in the past.

Operation and Maintenance

Depending on the use (i.e., intermittent use in the case of fire-fighting supplies or regular in the case of irrigation supplies) and water source used (e.g., seawater or wastewater), in the dual distribution system, regular testing of the system is recommended. The seawater-based system used in the U.S. Virgin Islands was tested daily in the past, but is now tested once a week. The pumps are turned on and a by-pass is used to allow the return of seawater to the sea to avoid pressurizing the distribution system. The pumps and engines are routinely serviced according to manufacturers' specifications.

Problems experienced in the operation and maintenance of this system include accidental damage to foot valves and standpipes. In the case of seawater systems, ships have been known to damage foot valves located in the harbor, and in the case of the distribution systems, vehicles frequently damage hydrants and standpipes, which then have to be replaced. In addition, foot valves require frequent servicing to remove fungal and other growths which can prevent their proper opening and closing and can make it impossible for the pumps to maintain their charge. On the landward side, regular inspection and maintenance of the standpipes and hydrants is required to remove debris from the openings of the hydrants and standpipes, which become clogged as a result of vandalism (persons pushing debris into the hydrant openings). It is also necessary to ensure that the pump engines arc supplied with adequate reserves of oil and fuel, and that the pumps and motors are properly lubricated for optimal operation.

Effectiveness of the Technology

This technology is highly effective. Seawater is as effective as potable water when used for fire-fighting purposes, but does not result in the drawdown of potable supplies. The system installed in Castries provides sufficient urban coverage and adequate supplies of water to fight most fires in the city. In contrast, public support for the dual distribution system in the U.S. Virgin Islands has diminished, making the system more prone to vandalism and less effective overall.

Suitability

The technology is suitable only in areas where a supply of raw water is available. This type of system is generally used near the coast where seawater is abundant, or in places where wastewater is readily available as a source of supply. It can also be utilized in areas that have rivers, streams, or other water sources but lack treatment facilities; in other words, in areas supplied with public water but having access to additional water sources that would otherwise go unutilized or underutilized.

Advantages and Disadvantages

This technology allows the use of cheaper sources of water for non-consumptive purposes, which may currently be served from more expensive, and limited, potable water supplies. If used to augment the regular distribution system, it makes more potable water available to the general public.

A dual distribution system requires that two distribution systems have to be installed, at essentially double the cost of a single system. Having non-potable water in a distribution system creates a potential to cross-contaminate the potable water system (while this is of limited concern in seawater systems, accidental consumption of non-potable water from wastewater-based systems could have serious consequences). Use of untreated seawater or wastewater to irrigate leafy vegetables could also threaten human health. Seawater can be highly corrosive to metal pipes, fittings, and appurtenances; it increases maintenance costs associated with distribution lines and affects toilet and other metal fixtures that come into contact with the water. If return flows enter the wastewater stream, the introduction of large volumes of seawater to treatment plants make sewage treatment more difficult since the salts can impair the effectiveness of activated sludge reactors or rotating biofilters, for example.

Further Development of the Technology

Development and use of non-corrosive materials, such as fiberglass, may make this technology more attractive, especially in cases where seawater is the principal source of non-potable water used in the dual distribution system. The use of alternative materials such as PVC in components such as foot valves might reduce potential for fungal growth and other growths that clog or damage the valves. There is also a great need for public awareness, among users, plumbers, and others, to minimize cross-connections and other potential sources of cross-contamination of the potable water supply.

Vocabulary

dual	双的,二重的	appurtenances	附属物
independently	独立地	activated sludge	活性污泥
non-potable	不适饮用的	biofilter	生物滤池
harbor	海港	fiberglass	玻璃丝
prone	倾向于	fungal	真菌的

Unit 8 Home Plumbing System

Plumbing system is pipes and fixtures installed in a building for the distribution and use of potable (drinkable) water and the removal of waterborne wastes. It is usually distinguished from water and sewage systems that serve a group of buildings or a city.

One of the problems of every civilization in which the population has been centralized in cities and towns has been the development of adequate plumbing systems. In certain parts of Europe the complex aqueducts built by the Romans to supply their cities with potable water can still be seen. However, the early systems built for the disposal of human wastes were less elaborate. Human wastes were often transported from the cities in carts or buckets or else discharged into an open, water-filled system of ditches that led from the city to a lake or stream.

Improvement in plumbing systems was very slow. Virtually no progress was made from the time of the Romans until the 19th century. The relatively primitive sanitation facilities were inadequate for the large, crowded population centre that sprang up during the Industrial Revolution, and outbreaks of typhoid fever and dysentery were often spread by the consumption of water contaminated with human wastes. Eventually these epidemics were curbed by the development of separate, underground water and sewage systems, which eliminated open sewage ditches. In addition, plumbing fixtures were designed to handle potable water and water-borne wastes within buildings.

The term plumbing fixture embraces not only showers, bathtubs, lavatory basins, and toilets but also such devices as washing machines, garbage-disposal units, hot-water heaters, dishwashers, and drinking fountains.

The water-carrying pipes and other materials used in a plumbing system must be strong, no corrosive, and durable enough to equal or exceed the expected life of the building in which they are installed. Toilets, urinals, and lavatories usually are made of stable porcelain or vitreous china, although they sometimes are made of glazed cast iron, steel, or stainless steel. Ordinary water pipes usually are made of steel, copper, brass, plastic, or other nontoxic material; and the most common materials for sewage pipes are cast iron, steel, copper, and asbestos cement.

Methods of water distribution vary. For towns and cities, municipally or privately owned water companies treat and purify water collected from wells, lakes, rivers, and ponds and distribute it to individual buildings. In rural areas water is commonly obtained directly from individual wells.

In most cities, water is forced through the distribution system by pumps, although, in rare instances, when the source of water is located in mountains or hills above a city, the pressure generated by gravity is sufficient to distribute water throughout the system. In other cases, water is pumped from the collection and purification facilities into elevated storage tanks and then allowed to flow throughout the system by gravity. But in most municipalities water is pumped directly through the system; elevated storage tanks may also be provided to serve as pressure-stabilization devices and as an auxiliary source in the event of pump failure or of a catastrophe, such as fire, that might require more water than the pumps or the water source are able to supply.

The pressure developed in the water-supply system and the friction generated by the water moving through the pipes are the two factors that limit both the height to which water can be distributed and the maximum flow rate available at any point in the system.

A building's system for waste disposal has two parts: the drainage system and the venting system. The drainage portion comprises pipes leading from various fixture drains to the central main, which is connected to the municipal or private sewage system. The venting system consists of pipes leading from an air inlet (usually on the building's roof) to various points within the drainage system; it protects the sanitary traps from siphoning or blowing by equalizing the pressure inside and outside the drainage system.

Sanitary fixture traps provide a water seal between the sewer pipes and the rooms in which plumbing fixtures are installed. The most commonly used sanitary trap is a U bend, or dip, installed in the drainpipe adjacent to the outlet of each fixture. A portion of the waste water discharged by the fixture is retained in the U, forming a seal that separates the fixture from the open drainpipes.

Vocabulary

plumbing	室内给排水系统	porcelain	瓷,瓷器
fixture	装置,设备	vitreous	玻璃质的
aqueduct	导水管,沟渠	asbestos	石棉,含石棉的
elaborate	复杂的,精致的	brass	铜器,黄铜
sanitation	卫生,卫生设施	auxiliary	辅助的,协助的
lavatory	洗脸盆	vent	通风口,排气道
urinal	小便器	siphon	虹吸管,虹吸

Reading Material A

Plumbing

One of most important systems developed to protect the health of man and to provide man with a better way of life has been the system of plumbing, which is the piping of potable water to its ultimate use and the draining away of waste materials to a variety of treatment processes. It is essential in our modern society to recognize that plumbing is to society what the circulatory system is to man. It is a system which must function efficiently to avoid outbreaks of epidemics and to avoid chemical pollution. Good health practices require that plumbing in a community be free of cross-connections, backflow connections, submerged inlets and poor venting. It also must transport a good quality of potable water in adequate quantities in order to service our modern society. One of the great difficulties that we face as a society is that older existing plumbing may deteriorate and may create health hazards; also, repairs of plumbing may be carried out in such a way that they will create direct health hazards.

Plumbing is the practice, materials, and fixtures used in installing, maintaining, and altering of pipes, appliances, and appurtenances, utilized for potable water supply, sanitary or storm drainage and venting systems. Plumbing does not include the drilling of water wells, installing water softening equipment, or sale of the manufacture or plumbing fixtures, appliances, equipment or hardware. Plumbing systems consist of an adequate potable water supply system, a safe adequate drainage system and ample fixtures and equipment.

Public health personnel have long been concerned with cross-connections, backflow connections and submerged inlets in plumbing systems and public drinking water supply distribution systems. These cross-connections make possible the contamination of potable water with nonpotable water or contaminated water. Although the probability of contamination of drinking water seems to be remote, a multitude of problems definitely exist. The only proper precaution is to eliminate all the possible links and channels where potable water may be polluted. Cross-connections exist when the individual installing the plumbing is not aware of the danger and may not realize that water can reverse its direction. In fact it may even go uphill. In addition, the valves may fail or may be carelessly left open. In order to combat this problem, installers must understand the hydraulic and pollution factors which can cause environmental health hazards. They must also know what types of standard backflow prevention devices and methods are utilized and how to obtain the materials and install them properly.

This text on plumbing is not meant to be an overall plumbing guide. It is not even meant to list all of the potential hazards. However, it should provide sufficient material and diagrams to help the environmental health practitioner have a better understanding of plumbing and its effects on health.

The current status of the plumbing problem is very difficult to ascertain since data is lacking in this area. However, it can be assumed that plumbing systems in many areas are rapidly deteriorating because of the age of the structure. Unfortunately, about the only new thing added to plumbing in the last 75 years has been the introduction of plastic pipes. As a result of this lack of change, many individuals fail to pay adequate attention to the enormous potential hazard of disease and injury due to microbiological, chemical or physical agents.

Vocabulary

circulatory	循环的,流通的	deteriorate	变质,损坏
outbreak	爆发,破裂	valve	阀,阀门
submerged	水面下的	uphill	上坡,向上
inlet	进水(气)口	ascertain	确定,探知

Reading Material B

Sewage System

Sewage system, the collection pipes and mains, treatment works, and discharge lines for the wastewater of a community.

Early civilizations often built drainage systems in urban areas to handle storm runoff. The Romans, especially, constructed elaborate systems that also drained wastewater from the public baths. During the European Middle Ages these systems fell into disrepair. As the populations of cities grew, disastrous epidemics of cholera and typhoid fever broke out, the result of ineffective segregation of sewage and drinking water. As the correlation between sewage and disease became apparent in the mid-19th century, steps were taken to treat wastewater. The concentration of population and the addition to sewage of manufacturing waste that occurred during the Industrial Revolution increased the need for effective sewage treatment.

Modern sewage systems fall under two categories: domestic and industrial sewers, and storm sewers. Sometimes a combined system provides only one network of pipes, sewer mains, and outfall sewers for all types of sewage. This type of system is less expensive to install in a district, but it has long been recognized that separate systems are best suited to modern metropolitan conditions. To avoid pollution, all waste water should pass through treatment plants, but it is uneconomical to build plants large enough to accept the enormously enlarged inflow from rainstorms and to treat sewage at the same time. In addition, a combined system must be made so large that it may not be able to provide adequate velocity for the dry-weather flow of the waste water alone. The preferred system provides separate sewers for human waste, which is then generally treated before discharge.

Sewer pipe is made of vitrified clay or concrete and its cross section is usually round. It is laid following street patterns, and access holes with metal covers are provided periodically for inspection and cleaning. Catch basins at street corners and along street gutters admit surface runoff of storm water and feed the storm sewers. Engineers determine the volume of sewage likely, the route of the system and the slope, or gradient, of the pipe to ensure an even flow by gravity that will not leave solids behind. In flat regions, pumping stations are sometimes needed. Under certain conditions, ventilating equipment is provided to remove corrosive gases.

Sewage treatment entails the removal of organic matter and is usually accomplished in two stages. In the first, or primary, stage sewage is first passed through large mesh screens to remove such large objects as wood, rags, and wire. It is then run through channels at a controlled velocity so that sand and ash grit is deposited on the bottom. After screening and grit removal, the sewage is passed into large tanks about 3 metres deep where many of the suspended solids (sludge) settle in a process called sedimentation. Two additional methods can supplement primary treatment. The Imhoff tank, developed by the German engineer Karl Imhoff, is a second compartment under the settling tank where solids are further decomposed by bacteria. Chemicals can also be added to the sewage to promote the coagulation of the finer suspended solids, so that they become heavy enough to settle in sedimentation.

The secondary treatment of sewage produces an effluent clean enough for discharge. The work of further purification is performed by microorganisms and bacterial slime, most commonly through the use of trickling and sand filters or the activated-sludge process. The organic matter remaining in sewage after solids have been removed is mainly in a dissolved state. The trickling filter is a bed of stones covered by a thin film of purifying slime through which sewage, sprayed from above, is allowed to trickle. It then runs onto a bed of sand that filters the water clean. The activated-sludge process utilizes sludge that has been allowed to breed microorganisms. The sludge is mixed with treated sewage and then aerated by jets of compressed air over a period of several hours; during this time the organic matter is oxidized by the microorganisms. Afterward the sewage is returned to settling tanks and then aerated a second time.

Sludge from both primary and secondary treatments is collected from the various tanks and hauled out to sea and dumped or buried in sanitary landfills. It may also be used, in liquid or dried-out form, as fertilizer. By placing it in digestion tanks heated to an optimal 35 ℃ it can be further decomposed to produce methane gas, which can be used to run the machinery of the treatment plant.

Vocabulary

segregation	分离,隔离	aerate	使暴露于空气中，使充满气体
metropolitan	大城市,主要城市	secondary treatment	二级处理
discharge	排放	digestion	消化
screen	格栅	decompose	分解，(使)腐烂
sludge	污泥	methane	甲烷
microorganism	微生物		
primary treatment	一级(初级)处理		

Unit 9 Water Treatment Processes

When people turn on a tap, clean, clear water runs out. But how does the local water supplier make sure the water is safe and pleasant to drink? The answer is likely to include some form of water treatment. Most urban communities collect water from a natural water body in the catchment, whether a stream, river, or underground aquifer. The water collected may then be stored in a reservoir for some time. Unless it is already of very high quality, it then undergoes various water treatment processes that remove any chemicals, organic substances or organisms that could be harmful to human health. The water is then delivered to the community through a network of mains and pipes called a distribution system.

Treatment designed to change polluted water to a potage water supply usually involves the following steps: sedimentation, with or without the use of a chemical flocculating agent, results in a settling out of the coarse particles that tend to carry down bacteria and similar smaller particles. The water may then be passed through fairly deep sand filters. The floc is caught in the sand and forms a rather efficient filter removing most of the organic particles and bacteria. After filtration, water is greatly improved in quality, appearance, and sanitary properties, but it is generally not entirely free from bacteria. Therefore the water is generally chlorinated.

Coagulation and Flocculation

In the coagulation/flocculation process, very fine suspended solids and colloidal particles are caused to come together to form larger particles that can be settled and filtered out of the water. Colloidal solids are particles with a diameter of less than 0.01 mm. These include fine silts, bacteria, color causing particles and viruses that might not settle for days, months or even years. Although individual particles cannot be seen with the naked eye, their combined effect is often seen as color or turbidity (cloudiness) in the water. These particles are small enough to pass through later treatment processes if not properly coagulated and flocculated. This could adversely affect not only the clarity of the water, but its taste and odor, as well as the effectiveness of chlorine disinfection. Therefore, a chemical (coagulant) is added to the raw water that causes fine particles to bind together (flocculate) to form larger, fluffy clusters (floc) that can be removed by settling and filtering.

In the coagulation process, the coagulant, Aluminum Sulfate or Iron Sulfate, is added to the incoming raw water. The water is stirred vigorously by an in-line flash mixer to assure quick, uniform dispersion of the alum. The alum reacts rapidly with the water's alkalinity (compounds similar to baking soda that contain carbonates, bicarbonates and hydroxides). This reaction produces a gelatinous (jelly-like) precipitate of Aluminum Hydroxide that entraps and absorbs impurities. At the same time, alum with a positive charge, neutralizes the negative charge common to natural particles, thus allowing them to come together. The floc that is first formed is very small particles called microfloc.

The water moves from the flash mixer to the flocculation basins which are baffled chambers containing horizontal paddle systems. The design of the flocculation basins results in a gentle, constant mixing of the microfloc formed during coagulation. This stirring provides maximum contact of the floc particles and promotes formation of larger and heavier floc. After 20 to 30 minutes, the floc particles are usually visible and will look like tiny tufts of cotton or wool, separated by clear water. Once the flow is of sufficient size and density to be settled, the water moves into the sedimentation basins.

Sedimentation

After the water leaves the flocculation basins, it enters the sedimentation or settling basins. Sedimentation is the removal of solids from water by gravity settling. Basins are designed to hold large volumes of water for several hours and to give a smooth, even flow. This design allows the velocity and turbulence of the water to be decreased to the point that the water will no longer transport the flocculated solids and they will settle to the bottom of the basin.

There are three basins of equal size that each hold about 5.3 million litres of water. The inlet side of each basin is 4.9 m deep and tapers up to a depth of 3.3 m at the outlet. Most of the solids settle out in the deep end; and as the water moves slowly through the basin, it gradually becomes clearer over a four to five hour period. The solids that collect as sludge are periodically removed, dewatered and disposed of by landfill.

At the end of the settling basin, hydrated lime, $Ca(OH)_2$, is added to the water to increase its pH. This is necessary because the alum that is added during the coagulation/flocculation process is an acidic salt that decreases the pH. After pH adjustment, chlorine is added for disinfection and to keep aquatic growths from becoming established on the filters during the next treatment process.

Filter

Settled water is pumped onto the top of the filters, leaving behind suspended matter as it passes downward through the filter. Some solids are larger than the pores or holes between the media grains, resulting in a straining action. Also, some suspended matter sticks to the surface (adsorbs) of the filter media or the previously deposited material. In addition, a little flocculation and sedimentation continues in the filter bed.

After a period of time, the accumulated solids begin to clog the filter and the filter is backwashed. This is reversal of the direction of the water flow through the filter. A rapid upward flow lifts the media and keeps it in suspension until the accumulated material can be washed out. Following filtration, the water flows into the storage tank (clearwell).

Water Stabilization

Some water supplies can become acidic or alkaline by dissolving or reacting with the material they are in contact with. This can cause piping systems and hot water services to corrode and cause dissolved metals to appear in the water.

For example, a common sign of copper corrosion is a bluish stain where a tap drips onto a surface. To prevent corrosion, many waters are chemically stabilized to a particular pH before distribution by adding lime and sometimes carbon dioxide.

Disinfection

Water is disinfected to kill any pathogens that may be present in the water supply and to prevent them from regrowing in the distribution systems. Without disinfection, the risk from waterborne disease is increased.

The two most common methods to kill microorganisms in the water supply are oxidation with chemicals such as chlorine or ozone or irradiation with ultra-violet (UV) radiation.

The most widely used chemical systems in Australia are chlorination, chloramination and ozonation. There is no ideal disinfectant and each has its own advantages and disadvantages. The goal is always to protect public health and the choice depends on the individual water quality and water supply system.

Other Treatments for Unusual Cases

While coagulation, often combined with filtration, will remove most of the troublesome contaminants from water, these processes do not usually remove all the material dissolved in the water. If the water contains undesirable impurities, additional treatment such as adsorption and oxidation may be required.

Adsorption is a form of chemical filtration that involves removing dissolved substances by chemically or physically binding them to the filter material. It is quite different from the similarly sounding process of absorption. In water treatment, specialized adsorbent materials such as activated carbon and ion exchange resins are used to remove certain soluble contaminants from water. One way of using activated carbon is to percolate water through a bed of carbon granules. Once the carbon is saturated with the contaminants, it needs to be replaced or regenerated by heating it to a high temperature. If water contamination occurs only occasionally, but can be detected by a regular monitoring program, a better approach is to add powdered activated carbon to a conventional coagulation/flocculation process when a problem arises. The saturated carbon is collected in the filters and then discarded with the normal sludge from the water treatment plant. This form of intermittent dosing is used widely in Australia where there are occasional problems with blue-green algal blooms, which can cause taste and odour problems, and can also be toxic.

Oxidation with chemicals such as ozone or chlorine dioxide, a common treatment technology in Europe, has appeared in Australia only recently. Strongly reactive chemicals such as ozone are used to disinfect water and to destroy soluble contaminants such as algal toxins, taste and odour compounds and, particularly in Europe, traces of pesticides.

Innovative water treatment technologies are being developed in Australia and overseas. Much of the Australian research is conducted within the Cooperative Research Centre for Water Quality and Treatment, where scientists, technologists and engineers are developing new water treatment technologies and improving existing ones. Their work will improve water quality for many communities and lower costs for households and businesses.

Vocabulary

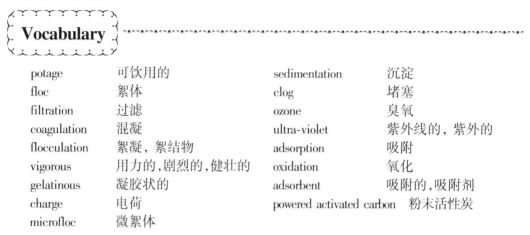

potage	可饮用的	sedimentation	沉淀
floc	絮体	clog	堵塞
filtration	过滤	ozone	臭氧
coagulation	混凝	ultra-violet	紫外线的，紫外的
flocculation	絮凝，絮结物	adsorption	吸附
vigorous	用力的，剧烈的，健壮的	oxidation	氧化
gelatinous	凝胶状的	adsorbent	吸附的，吸附剂
charge	电荷	powered activated carbon	粉末活性炭
microfloc	微絮体		

Reading Material A

The History of Drinking Water Treatment

Ancient civilizations established themselves around water sources. While the importance of ample water quantity for drinking and other purposes was apparent to our ancestors, an understanding of drinking water quality was not well known or documented. Although historical records have long mentioned aesthetic problems (an unpleasant appearance, taste or smell) with regard to drinking water, it took thousands of years for people to recognize that their senses alone were not accurate judges of water quality.

Water treatment originally focused on improving the aesthetic qualities of drinking water. Methods to improve the taste and odor of drinking water were recorded as early as 4000 B.C. Ancient Sanskrit and Greek writings recommended water treatment methods such as filtering through charcoal, exposing to sunlight, boiling, and straining. Visible cloudiness (later termed turbidity) was the driving force behind the earliest water treatments, as many source waters contained particles that had an objectionable taste and appearance. To clarify water, the Egyptians reportedly used the chemical alum as early as 1500 B.C. to cause suspended particles to settle out of water. During the 1700s, filtration was established as an effective means of removing particles from water, although the degree of clarity achieved was not measurable at that time. By the early 1800s, slow sand filtration was beginning to be used regularly in Europe.

During the mid to late 1800s, scientists gained a greater understanding of the sources and effects of drinking water contaminants, especially those that were not visible to the naked eye. In 1855, epidemiologist Dr. John Snow proved that cholera was a waterborne disease by linking an outbreak of illness in London to a public well that was contaminated by sewage. In the late 1880s, Louis Pasteur demonstrated the "germ theory" of disease, which explained how microscopic organisms (microbes) could transmit disease through media like water.

During the late nineteenth and early twentieth centuries, concerns regarding drinking water quality continued to focus mostly on disease-causing microbes (pathogens) in public water supplies. Scientists discovered that turbidity was not only an aesthetic problem; particles in source water, such as fecal matter, could harbor pathogens. As a result, the design of most drinking water treatment systems built in the U. S. during the early 1900s was driven by the need to reduce turbidity, thereby removing microbial contaminants that were causing typhoid, dysentery, and cholera epidemics. To reduce turbidity, some water systems in U.S. cities (such as Philadelphia) began to use slow sand filtration.

While filtration was a fairly effective treatment method for reducing turbidity, it was disinfectants like chlorine that played the largest role in reducing the number of waterborne disease outbreaks in the early 1900s. In 1908, chlorine was used for the first time as a primary disinfectant of drinking water in Jersey City, New Jersey. The use of other disinfectants such as ozone also began in Europe around this time, but were not employed in the U.S. until several decades later.

Federal regulation of drinking water quality began in 1914, when the U.S. Public Health Service set standards for the bacteriological quality of drinking water. The standards applied only to water systems which provided drinking water to interstate carriers like ships and trains, and only applied to contaminants capable of causing contagious disease. The Public Health Service revised and expanded these standards in 1925, 1946, and 1962. The 1962 standards, regulating 28 substances, were the most comprehensive federal drinking water standards in existence before the Safe Drinking Water Act of 1974. With minor modifications, all 50 states adopted the Public Health Service standards either as regulations or as guidelines for all of the public water systems in their jurisdiction.

By the late 1960s it became apparent that the aesthetic problems, pathogens, and chemicals identified by the Public Health Service were not the only drinking water quality concerns. Industrial and agricultural advances and the creation of new man-made chemicals also had negative impacts on the environment and public health. Many of these new chemicals were finding their way into water supplies through factory discharges, street and farm field runoff, and leaking underground storage and disposal tanks. Although treatment techniques such as aeration, flocculation, and granular activated carbon adsorption (for removal of organic contaminants) existed at the time, they were either underutilized by water systems or ineffective at removing some new contaminants.

Health concerns spurred the federal government to conduct several studies on the nation's drinking water supply. One of the most telling was a water system survey conducted by the Public Health Service in 1969 which showed that only 60 percent of the systems surveyed delivered water that met all the Public Health Service standards. Over half of the treatment facilities surveyed had major deficiencies involving disinfection, clarification, or pressure in the distribution system (the pipes that carry water from the treatment plant to buildings), or combinations of these deficiencies. Small systems, especially those with fewer than 500 customers, had the most deficiencies. A study in 1972 found 36 chemicals in treated water taken from treatment plants that drew water from the Mississippi River in Louisiana. As a result of these and other studies, new legislative proposals for a federal safe drinking water law were introduced and debated in Congress in 1973.

Chemical contamination of water supplies was only one of many environmental and health issues that gained the attention of Congress and the public in the early 1970s. This increased awareness eventually led to the passage of several federal environmental and health laws, one of which was the Safe Drinking Water Act of 1974. That law, with significant amendments in 1986 and 1996, is administered today by the U. S. Environmental Protection Agency's Office of Ground Water and Drinking Water (EPA) and its partners. Since the passage of the original Safe Drinking Water Act, the number of water systems applying some type of treatment to their water has increased. According to several EPA surveys, from 1976 to 1995, the percentage of small and medium community water systems (systems serving people year-round) that treat their water has steadily increased. For example, in 1976 only 33 percent of systems serving fewer than 100 people provided treatment. By 1995, that number had risen to 69 percent.

Since their establishment in the early 1900s, most large urban systems have always provided some treatment, as they draw their water from surface sources (rivers, lakes, and reservoirs) which are more susceptible to pollution. Larger systems also have the customer base to provide the funds needed to install and improve treatment equipment. Because distribution systems have extended to serve a growing population (as people have moved from concentrated urban areas to more suburban areas), additional disinfection has been required to keep water safe until it is delivered to all customers.

Today, filtration and chlorination remain effective treatment techniques for protecting U.S. water supplies from harmful microbes, although additional advances in disinfection have been made over the years. In the 1970s and 1980s, improvements were made in membrane development for reverse osmosis filtration and other treatment techniques such as ozonation. Some treatment advancements have been driven by the discovery of chlorine-resistant pathogens in drinking water that can cause illnesses like hepatitis, gastroenteritis, Legionnaire's Disease, and cryptosporidiosis. Other advancements resulted from the need to remove more and more chemicals found in sources of drinking water. According to a 1995 EPA survey, approximately 64 percent of community ground water and surface water systems disinfect their water with chlorine. Almost all of the remaining surface water systems, and some of the remaining ground water systems, use another type of disinfectant, such as ozone or chloramine.

Many of the treatment techniques used today by drinking water plants include methods that have been used for hundreds and even thousands of years. However, newer treatment techniques (e.g., reverse osmosis and granular activated carbon) are also being employed by some modern drinking water plants.

Recently, the Centers for Disease Control and Prevention and the National Academy of Engineering named water treatment as one of the most significant public health advancements of the 20th Century. Moreover, the number of treatment techniques, and combinations of techniques, developed is expected to increase with time as more complex contaminants are discovered and regulated. It is also expected that the number of systems employing these techniques will increase due to the recent creation of a multi-billion dollar state revolving loan fund that will help water systems, especially those serving small and disadvantaged communities, upgrade or install new treatment facilities.

Vocabulary

aesthetic	美学的,审美的	dysentery	痢疾
epidemiologist	流行病学家	contagious	传染性的
transmit	传输,传送	susceptible	易受影响的，易感动的，易感染的
fecal	排泄物的		
harbor	挟带,隐藏	reverse osmosis	反渗透

Reading Material B

Storm Water

Storm water is a widespread problem. Storm water carries contaminants from sites into the collection systems. These contaminants enter water supplies, create turbidity, and increase bacterial contamination. Storm water dilutes the concentration of food in treatment plants, causing problems with microbes, and can overwhelm plants and cause excessive discharge.

Storm water is a type of wastewater which consists of water running off roofs, streets, and other surfaces. Storm water is usually kept separate from other types of wastewater since, in most cases, storm water does not require treatment. If storm water is mixed with sewage, then the entirety of the mixture will have to be treated in a sewage treatment plant, a process which will be much more costly than if the sewage was treated alone.

Diluting sewage with storm water makes treatment costly and difficult for a variety of reasons. The addition of storm water results in a greater volume of wastewater. The diluted sewage will have a different concentration of wastes, so the amounts of sewage treatment organisms will have to be adjusted. And, since storm water typically enters sewage lines in large quantities at infrequent intervals, storm water which is allowed to combine with sewage can wash out a wastewater treatment plant.

For all of these reasons, storm water and sewage are best kept separate during treatment. In a few cases, storm water requires treatment even when it has not been mixed with sewage. The EPA's National Pollutant Discharge Elimination System (NPDES) regulates industrial storm water by requiring industries to acquire a permit for every point source which may discharge pollutants into natural waterways.

Construction sites are another potential source of problematic storm water. Disturbed soil will erode during storms so that the resulting storm water is full of sediments. The DEQ requires that anyone disturbing at least 10,000 sq. ft. of soil must develop a water plan. This plan may involve sedimentation ponds which retain storm water and allow the sediments to settle out, thus releasing the storm water over a longer period of time and minimizing the impact downstream. The plan may also include cover crops, briars, and weeds planted on the bare soil. These plants will help reclaim the site by slowing the velocity of storm water running over the site and preventing it from picking up sediments from the ground.

Storm water running over mines will often pick up metals as well as sediments. In this case, the pH of the sedimentation ponds is raised to approximately 9 to precipitate out most of the metals and other contaminants that may have been picked up by the water. In most cases, though, local utilities do not need to treat storm water. Natural processes clean storm water, removing sediments, ammonia, and microorganisms.

Erosion is also treated by nature. Quickly moving water erodes stream banks, causing the rocks higher on the bank to roll down into the middle of the stream. Faced with the obstacle of large rocks in the stream bed, the water naturally slows down and does not cause as much erosion. Bare ground is quickly covered by plants which prevent storm water from picking up sediments from the soil. As you can see, both man and nature have devised many ways of minimizing erosion and keeping water clean.

Vocabulary

dilute	冲淡，变淡，变弱，稀释	ammonia	氨，氨水
erosion	腐蚀，侵蚀	downstream	下游的

Unit 10 Mixing

Chemical solutions for flocculation and coagulation are added prior to the settling and clarification basins. The designers of earlier water treatment plants had litter appreciation of the importance of rapid mixing. Chemicals were added to the influent water to mix naturally with no attempt to mechanically accelerate to the process. When chemical reaction kinetics were studied, it was realized that flocs form almost instantaneously when chemicals first come into contact with the water, and the importance of rapid mixing was appreciated. Much of the research has been done by Camp. Some of his conclusions are as follows:

1. The floc volume concentration and size distribution is determined during flocculation by the mean velocity gradient and time, and both the concentration and the size of floc may be varied over a wide range by changes in velocity gradient and time.

2. After flocculation is complete, at a particular velocity gradient, continued mixing at the same velocity gradient results in little change in floc volume concentration and size distribution.

3. Rapid mixing at sufficiently high velocity gradients of floc already formed will dispense such floc into colloidal particles.

The problems inherent in mixing a small quantity of chemical solution into a large quantity of water are discussed late. Complete mixing of the chemical into the water before the two react is strongly emphasized. If the mixing is too slow some parts of the water will be subjected to chemical overdose while other parts, the large majority of the flow, will have no chemical dosage at all. If agitation is too violent, the flocs initially formed will be broken into colloids. This can be worse than the initial condition, since colloids may not recoagulate and are too small to settle out. No clear-cut guidelines appear to exist for determining the optimum power dissipation, design of mixer, or detention time required to disperse chemicals into a flow of water. However, detention times of 20 seconds or less are not uncommon, and mixing units providing 735.5 to 1471 W for each cubic foot per second of flow rate are not unreasonable.

Much of the modern design date is in terms of velocity gradients. To illustrate the meaning of this term, if two particles of fluid in a tank are 0.03 m apart and one is moving with a speed of 0.3 m/s relative to the other, the velocity gradient between them is 10. The velocity gradient is therefore the relative velocities of two particles divided by their distance apart. Velocity gradients can be calculated from the following equations.

For baffled basins: $G = \sqrt{\frac{62.4H}{\mu T}}$

For mechanical agitation: $G = \sqrt{\frac{62.4P}{V\mu}}$

Where:

G—Velocity gradient has the dimensions of velocity/distance which is m/sec ÷ m and is equal to (sec^{-1});

H—Head loss due to friction in feet;

μ—Viscosity in $\frac{1b\ sec}{ft^2}$ units which is equivalent to centipoises $\times 2.088 \times 10^{-5}$;

T—Detention time in seconds;

V—Volume of basin in cubic feet;

P—Water horsepower.

Camp studied the impact of rapid mixing on floc formation. He reported that mixing at G values of 500 sec^{-1} to 1 000 sec^{-1} for about 2 minutes produced essentially complete flocculation and that prolonged rapid mixing at this intensity accomplished practically nothing more. For rapid mixing with small G values, a 2 minute period was insufficient to complete the flocculation; on the other hand, mixing at excessively high G values (10 000 sec^{-1} or more) for as long a time as 2 minutes seemed to substantially retard floc formation.

The total number of particle collisions is proportional to GT. Mixing and flocculation are more efficient when imparted as controlled turbulence, caused by propeller mixers and baffles, rather than as a general rotation of the mass of water or a flow along a conduit.

Other authorities have found that the degree of agitation produced is of greater importance than the time over which the mixing continues and, if a special mixing chamber is provided, adequate dispersal of chemicals can be obtained in a period as short as 30 seconds to 1 minute if sufficient power is provided for the purpose. The power required for efficient mixing is given as 228 to 457 mm head of water which is approximately equivalent to 184 to 368 W hp per 3 785 litres per minute.

Variable-speed driven agitations capable of a wide variety of velocity gradients should be installed. The enormously improved results in treated water quality, and the reduction in chemical dosage when optimum mixing is achieved, justifies the additional costs. It is not too difficult to design a suitable process to mix chemicals into natural water. The difficult problem is to add additional chemicals to modify already formed flocs, for example, the addition of ployelectrolytes as filter aid after the flocculation and clarifier processes. In this area of technology we do not have good answers, as rapid mixing for dispersion of the secondary flocculant or filter aid would break the already formed floc into colloids. A considerable amount of research has been devoted to the subject of flash mixing and velocity gradient.

Vocabulary

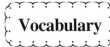

clarification	澄清,净化	agitation	搅动
rapid mixing	快速混合	head loss	水头损失
accelerate	加速,促进	turbulence	湍流,紊流
velocity gradient	速度梯度	insufficient	不足的,不充分的
colloidal	胶状的,胶质的	excessively	过分地,非常地
colloid	胶体	ployelectrolyte	聚合电解质,聚合电解液

Reading Material A

Coagulation and Flocculation

All waters, especially surface waters, contain both dissolved and suspended particles. Coagulation and flocculation processes are used to separate the suspended solids portion from the water.

The suspended particles vary considerably in source, composition charge, particle size, shape, and density. Correct application of coagulation and flocculation processes and selection of the coagulants depend upon understanding the interaction between these factors. The small particles are stabilized (kept in suspension) by the action of physical forces on the particles themselves. One of the forces playing a dominant role in stabilization results from the surface charge present on the particles. Most solids suspended in water possess a negative charge and, since they have the same type of surface charge, repel each other when they come close together. Therefore, they will remain in suspension rather than clump together and settle out of the water.

Coagulation and flocculation occur in successive steps intended to overcome the forces stabilizing the suspended particles, allowing particle collision and growth of floc. If one step is incomplete, the following step will be unsuccessful.

Coagulation

The first step destabilizes the particle's charges. Coagulants with charges opposite those of the suspended solids are added to the water to neutralize the negative charges on dispersed non-settlable solids such as clay and color-producing organic substances.

Once the charge is neutralized, the small suspended particles are capable of sticking together. The slightly larger particles, formed through this process and called microflocs, are not visible to the naked eye. The water surrounding the newly formed microflocs should be clear. If it is not, all the particles' charges have not been neutralized, and coagulation has not been carried to completion. More coagulant may need to be added.

A high-energy, rapid-mix to properly disperse the coagulant and promote particle collisions is needed to achieve good coagulation. Over-mixing does not affect coagulation, but insufficient mixing will leave this step incomplete. Coagulants should be added where sufficient mixing will occur. Proper contact time in the rapid-mix chamber is typically 1 to 3 minutes.

Flocculation

Following the first step of coagulation, a second process called flocculation occurs. Flocculation, a gentle mixing stage, increases the particle size from submicroscopic microfloc to visible suspended particles.

The microflocs are brought into contact with each other through the process of slow mixing. Collisions of the microfloc particles cause them to bond to produce larger, visible flocs called pinflocs. The floc size continues to build through additional collisions and interaction with inorganic polymers formed by the coagulant or with organic polymers added. Macroflocs are formed. High molecular weight polymers, called coagulant aids, may be added during this step to help bridge, bind, and strengthen the floc, add weight, and increase settling rate. Once the floc has reached it optimum size and strength, the water is ready for the sedimentation process.

Design contact times for flocculation range from 15 or 20 minutes to an hour or more.

Coagulant Selection

The choice of coagulant chemical depends upon the nature of the suspended solid to be removed, the raw water conditions, the facility design, and the cost of the amount of chemical necessary to produce the desired result.

Final selection of the coagulant (or coagulants) should be made following thorough jar testing and plant scale evaluation. Considerations must be given to required effluent quality, effect upon down stream treatment process performance, cost, method and cost of sludge handling and disposal, and net overall cost at the dose required for effective treatment.

Inorganic Coagulants Inorganic coagulants such as aluminum and iron salts are the most commonly used. When added to the water, they furnish highly charged ions to neutralize the suspended particles. The inorganic hydroxides formed produce short polymer chains which enhance microfloc formation.

Inorganic coagulants usually offer the lowest price per pound, are widely available, and when properly applied, are quite effective in removing most suspended solids. They are also capable of removing a portion of the organic precursors which may combine with chlorine to form disinfection by-products. They produce large volumes of floc which can entrap bacteria as they settle. However, they may alter the pH of the water since they consume alkalinity. When applied in a lime soda ash softening process, alum and iron salts generate demand for lime and soda ash. They require corrosion-resistant storage and feed equipment. The large volumes of settled floc must be disposed of in an environmentally acceptable manner.

Polymers Polymers—long-chained, high-molecular-weight, organic chemicals—are becoming more widely used, especially as coagulant aids together with the regular inorganic coagulants. Anionic (negatively charged) polymers are often used with metal coagulants. Low-to-medium weight, positively charged (cationic) polymers may be used alone or in combination with the aluminum and iron type coagulants to attract the suspended solids and neutralize their surface charge. The manufacturer can produce a wide range of products that meet a variety of source-water conditions by controlling the amount and type of charge and relative molecular weight of the polymer.

Polymers are effective over a wider pH range than inorganic coagulants. They can be applied at lower doses, and they do not consume alkalinity. They produce smaller volumes of more concentrated, rapidly settling floc. The floc formed from use of a properly selected polymer will be more resistant to shear, resulting in less carryover and a cleaner effluent.

Vocabulary

dissolved	溶解的	pinfloc	针尖絮体
suspended	悬浮的	molecular	分子,分子的
density	密度	polymer	聚合物
stabilize	稳定	coagulant aids	助凝剂
dominant	主导的,主要的	hydroxide	氢氧化物
negative	负的	soften	软化
destabilize	脱稳	cationic	阳离子
neutralize	中和,使中立		

Reading Material B

Optimizing Coagulation

For ages, water treatment plant operators have utilized jar tests for determining the best dosage of coagulant to achieve the best settling in the basins and jar testing has become the standard by which all other methods are judged. In the early 1970's, zeta potential monitoring became standardized and was made available at a reasonable price. The company, "Zeta-Meter" has since become a common name in water treatment plants. Shortly thereafter, another method, called streaming current monitoring was made available, and has achieved some success. The streaming current monitor will be discussed briefly here.

The streaming current monitor (SCM) uses an electric sensor to determine when charge neutralization has been reached in a suspension. The theory of operation is similar to that of the zeta-meter in that the charge-measuring device is based on zeta (elctrophoretic mobility). This device measures the net ionic and surface charges of colloids in suspension between two electrodes. A piston moves the water back and forth in the chamber and positive and negative charges are moved downstream to the electrodes, producing a streaming current. The streaming current amplitude and polarity is a function of the sampling location and the type of coagulant used.

If the correct coagulant dose is used (the correct amount of Fe^{3+} or Al^{3+} or other coagulant with positive valence), to neutralize the number of negative ions in the water, the SCM will read "0.0" indicating that charge neutralization has occurred. The SCM is not typical of most in-line instruments in that it is not "plug and play" and must be calibrated to a baseline. This baseline will vary from one raw water to another and from one treatment facility to another. To establish a baseline, tests should be conducted to establish a range of SCM readings when the plant is operating at its optimal coagulant dosage. Test results should include the effects of seasonal water quality changes. It is preferable to perform jar tests to acquire optimal coagulant dosage and record the SCM reading immediately after coagulation.

Through testing of the SCM in variable qualities of water, it appears that pH and alkalinity have the greatest effect on the instrument's usefulness. During high pH coagulation, the SCM's response time slows dramatically and at pH values above 8, the unit may be unresponsive to dosage changes. However, the intent of the SCM was to monitor water when attempting to coagulate by charge neutralization. Generally, at pH values of 8 or higher, sweep floc method is used, as it is nearly impossible to coagulate with charge neutralization. Although pH and alkalinity slow response time and require more frequent cleaning and de-calcification of the electrodes, the unit may still prove useful.

Normally, in cold, clear water, assuming the SCM has proven valuable, the operator will attempt to keep the SCM reading slightly below zero and as the temperature and organic particulates increase, raise to the reading through adjustment of coagulant feed to a point slightly higher than zero. Of course, these numbers will vary among different facilities. Periodically reconfirming the baseline will help validate the reliability of the instrument.

In conclusion, the SCM is a useful tool in coagulation optimization for treatment facilities wishing to achieve charge neutralization. It can record slight changes in chemical dosages thus saving costs while increasing reduction of TOC. It can also alert an operator to a feed system failure. The SCM requires an understanding of mixing kinetics and chemical reactions involved in coagulation. The usefulness of the instrument will depend largely on the quality of the water being coagulated, mainly pH and alkalinity. Higher pH values and alkalinity tend to slow or even stop the response of the instrument. SCM appears from studies to be most valuable when coagulation pH values are 7.0 or lower. The ability to study trends of the SCM under varying conditions and the establishment of a baseline should be the foremost considerations in deciding whether the instrument would be practical for an individual utility.

Vocabulary

jar test	杯罐试验	steaming current	流动电流
dosage	药剂量	neutralization	中和,中立化
potential	电位,势	reaction	反应

Unit 11 Coagulation and Flocculation

Impurities in water often cause the water to appear turbid or be colored. These impurities include suspended and colloidal materials and soluble substances. Because the density of many of these particles is only slightly greater than the density of water, agglomeration or aggregation of particles into a larger floc is a necessary step for their removal by sedimentation. The process that combines the particles into larger flocs is called coagulation.

Impurities that can be removed by coagulation include turbidity, bacteria, algae, color, organic compounds, oxidized iron and manganese, calcium carbonate, and clay particles. Clays are a large part of natural turbidity in raw waters, but are not directly responsible for harmful effects to humans. However, there is some evidence that clays affect human health indirectly through adsorption, transport, and release of inorganic and organic toxic constituents, viruses, and bacteria. Removal efficiencies of clay particles in water treatment are not normally monitored, although several laboratory studies have demonstrated the effectiveness of alum coagulation on clay suspensions.

Coagulation has been shown to be effective for the removal of color and other organic constituents in water. Laboratory studies have demonstrated the removal of these constituents using iron and aluminum salts. Humic acids were readily removed using these salts, although a large fraction of the fulvic acids was not removed. Because humic acids react with chlorine in the formation of halomethanes, their removal in the coagulation process is an important step in limiting the production of potential carcinogens.

The aggregation of colloidal particles takes place in two separate and distinct phases. First, the repulsion force between particles must be overcome, a step that requires that the particles be destabilized; and second, contact between the destabilized particles must be induced so that aggregation can occur. The destabilization step typically is achieved through the addition of chemicals, followed by thorough blending in rapid mix tanks. The aggregation step is accomplished through gentle stirring in flocculation tanks. A representation of the coagulation process is shown in Fig. 11.1.

It is useful to define the terms used in connection with the coagulation process. Coagulation is defined as the process that causes a reduction of repulsion forces between particles of the neutralization of the charges on particles. Flocculation is defined as the aggregation of particles into larger elements. The coagulation (destabilization) step is virtually instantaneous following addition of the coagulant, while the flocculation (transport) step requires more time for development of large flocs.

Historically, two theories have been advanced to explain the coagulation process of colloidal systems:

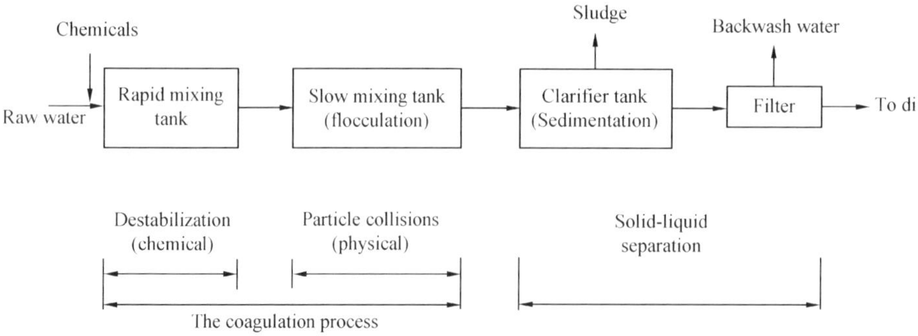

Fig. 11.1 Schematic diagram of a coagulation process

- Chemical theory;
- Physical or double-layer theory.

The former theory presumes that colloids acquire electrical charges on their surfaces by ionization of chemicals present at the surface, and that coagulation, or destabilization, occurs through specific chemical interactions between the coagulation and the colloids. Accordingly, this theory presumes that the coagulation of colloids is a result of the precipitation of insoluble complexes that are formed by chemical reactions.

The second theory is based on the presence of physical factors, such as electrical double layers surrounding the colloidal particles in the solution and counterion adsorption. Destabilization requires a reduction in the electric potential between the fixed layer of counterions and the bulk of the liquid. This electric, or zeta, potential can be estimated by observing the movement of microscopically visible particles in an electric field. This theory is presented in more detail below. These two mechanisms are not mutually exclusive. Both theories are used to explain the process of coagulation in treatment systems containing a heterogeneous mixture of colloids.

Vocabulary

impurity	杂质	instantaneous	即时的,瞬间的
agglomeration	结块,凝聚	double-layer	双电层
aggregation	聚合,聚集	electric field	电场
calcium carbonate	碳酸钙	exclusive	排外的
clay	黏土	heterogeneous	不同类的
fulvic acid	富里酸	adsorption	吸附
humic acid	腐殖酸	presume	假设,假定
halomethane	卤甲烷	microscopical	显微镜的
carcinogen	致癌物质	mechanism	机理
repulsion	排斥	mutually	相互地

Reading Material A

Theories of Coagulation

The Electrical Double Layer

When a colloidal particle is immersed solution, electrical charges develop at the particle-water interface. However, a colloidal dispersion does not have a net electrical charge. For electroneutrality to exist, the charges on the colloids must be counterbalanced by ions of opposite charge in the solution. The ions involved in establishing the electroneutrality are arranged in an electrical double layer. The concept of the electrical double layer was proposed initially by Helmholtz and later modified and improved by Gouy, Chapman, and Stern.

Double-layer model can be used to describe the electrical potential in the vicinity of a colloid particle. A portion of the counterions remain in a compact ("Stern") layer on the colloid surface. The remainder of the counterions extend into bulk of the solution, and constitute the diffuse ("Gouy-Chapman") layer. The effective thickness of the double layer is influenced significantly by the ionic concentration of the solution, but relatively little by the size of the colloid.

The electrical potential created by the surface charges will attract counterions toward the colloidal particles. The closest approach of the counterions to the particle is limited by the size of the ions. Stern proposed that the center of the closest counterions is separated from the surface charge by a layer of thickness, which represents the Stern layer. The electrical potential drops linearly across this layer. Beyond the Stern layer, in the diffuse layer, the electrical potential decreases exponentially with distance from the particle.

The magnitude of the charge on a colloid cannot be measured directly, but the value of the potential at some distance from the colloid can be computed. This potential, termed the zeta potential, can be computed by several techniques, such as electrophoresis, electroosmosis and streaming potential. Most often, the electrophoretic mobility of the colloidal particles is used to compute the zeta potential, by observing the particle mobility through a microscope.

Coagulation Mechanisms

Coagulation can be accomplished through any of four different mechanisms:

- Double layer compression;
- Adsorption and charge neutralization;
- Enmeshment by a precipitate (sweep-floc coagulation);
- Adsorption and interparticle bridging.

Double Layer Compression. This mechanism relies on compressing the diffuse layer surrounding a colloid. This is accomplished by increasing the ionic strength of the solution through the addition of an indifferent electrolyte (neutral salt). The explanation for this phenomenon lies in the Schulze-Hardy rule for anions, which was based on Schulze's work on the coagulating power of cations. He noted that the coagulating power increases in the ratio of 1:10:1 000 as the valency increases from 1 to 2 to 3. The Schulze-Hardy rule for anions is: the coagulating power of a salt is determined by valency of one of its ions. The prepotent ion is either the negative or positive ion, according to whether the colloidal particles more down or up the potential gradient. The coagulating ion is always of the opposite electrical sign to the particle.

This rule is valid for indifferent electrolytes, which are those that do not react with the solution. If such an electrolyte is added to a colloidal dispersion, the particle's surface charge will remain the same, but the added electrolyte will increase the charge density in the diffuse layer. This results in a smaller diffuse-layer volume being required to neutralize the surface charge. In other words, the diffuse layer is "compressed" toward the particle surface, reducing the thickness of the layer. At high electrolyte concentrations, particle aggregation can occur rapidly.

Adsorption and Charge Neutralization. The energy involved in an electrostatic interaction having a 100-millivolt potential difference across the diffuse layer between a colloidal particle and a monovalent coagulant ion is only about 9.66 kJ/mol. This compares with covalent bond energies in the range of 210 to 420 kJ/mol. Based on these facts, it is apparent that some coagulants can overwhelm the electrostatic effects and can be adsorbed on the surface of the colloid. If the coagulant carries a charge opposite to that of the colloid, a reduction in the zeta potential will occur, resulting in destabilization of the colloid. This process is quite different from the double layer compression mechanism described above.

The hydrolyzed species of Al(III) and Fe(III) can cause coagulation by adsorption. However, at higher doses of Al(III) coagulation is caused by enmeshment of the colloidal particles in a precipitate of aluminum hydroxide. This aspect is discussed in the next section. Destabization by adsorption is stoichiometric. Therefore, the requited coagulant dosage increases with increasing concentrations of colloids in the solution.

Enmeshment by a Precipitate (Sweep-Floc Coagulation). The addition of certain metal salts, oxides, or hydroxides to water in high enough dosages results in the rapid formation of precipitates. These precipitates enmesh the suspended colloidal particles as they settle. Coagulants such as aluminum sulfate ($Al_2(SO_4)_3$), ferric chloride ($FeCl_3$), and lime (CaO or $Ca(OH)_2$) are frequently used as coagulants to form the precipitates of $Al_2(SO_4)_3(s)$, $Fe(OH)_3(s)$ and $CaCO_3(s)$. The removal of colloids by this method has been termed sweep-floc coagulation.

This process can be enhanced when the colloidal particles themselves serve as nuclei for the formation of the precipitate. Therefore, the rate of precipitation increases with an increasing concentration of colloidal particles (turbidity) in the solution. Packham reported the inverse relationship between the optimum coagulant dose and the concentration of the colloids to be removed. Benefield explained this phenomenon as follows:

At low colloidal concentrations a large excess of coagulant is required to produce a large amount of precipitate that will enmesh the relatively few colloidal particles as it settles. At high colloidal concentrations, coagulation will occur at a lower chemical dosage because the colloids serve as nuclei to enhance precipitate formation.

This method of coagulation does not depend upon charge neutralization, so an optimum coagulant dose does not necessarily correspond to minimum zeta potential. However, an optimum pH does exist for each coagulant.

Destabilization by Interparticle Bridging. Synthetic polymeric compounds have been shown to be effective coagulants for the destabilization of colloids in water. These coagulants can be characterized as having large molecular sizes, and multiple electrical charges along a molecular chain of carbon atoms. Both positive (cationic) and negative (anionic) polymers are capable of destabilizing negatively charged colloidal particles. Surprisingly, the most economical destabilization process is often obtained using anionic polymers. The mechanisms already described cannot be used to describe this phenomenon, although a generally accepted chemical bridging theory/model has been developed that can explain the unusual reaction associated with synthetic polymer compounds.

The chemical bridging theory may be explained as follows. The simplest form of bridging proposes that a polymer molecule will attach to a colloidal particle at one or more sites. Colloidal attachment is postulated to occur as a result of coulombic attraction if the charges are of opposite charge, or from ion exchange, hydrogen bonding, or van der Waal's forces. The second reaction is the remaining length of the polymer molecule from the first reaction extends out into the bulk of the solution. If a second particle having some vacant adsorption sites contacts the extended polymer, attachment can occur to form a chemical bridge. The polymer, then, serves as the bridge. However, if the extended polymer molecule dose not contact another particle, it can fold back on itself and adsorb on the remaining sites of the original particle. In this event, the polymer is no longer capable of serving as a bridge, and in fact, it restabilizes the original particle.

Vocabulary

electroneutrality	电中和	compressed	压缩的
counterbalance	使平均，使平衡，弥补	monovalent	单价的
counterion	反平衡离子	stoichiometric	化学当量的,化学计算的
electroosmosis	电渗	coulombic	库仑的,库仑定律的
valency	化合价	van der Waal's force	范德华力
enmeshment	网捕	anionic	阴离子

Reading Material B

Water Treatment Chemicals

Potassium Permanganate

Potassium permanganate ($KMnO_4$) is a powerful oxidizing agent that is normally added to the incoming raw water. Potassium permanganate aids in the removal of iron and manganese and objectionable tastes and odors. Because the raw water is drawn from below the surface of the lake, dissolved oxygen is low. When the oxygen level is low, iron and manganese are kept in solution. If they are not removed during the treatment process, these metals will cause stains on consumers' laundry and plumbing fixtures. When iron and manganese are oxidized, they become insoluble in water and can then be removed by sedimentation and filtration.

Potassium permanganate is also used to oxidize the organic contaminants that cause taste and odor problems. In the past, chlorine was used for these problems as well as for iron and manganese control. However, research has shown that chlorine plus organic material can intensify some unpleasant tastes and odors and can produce chlorination by-products such as trihalomethanes. By using potassium permanganate instead of chlorine for pre-treatment, chlorine can be added later in the treatment process and used only for disinfection. Consumers thus receive a healthier and more aesthetically pleasing drinking water.

Alum

Aluminum Sulfate is $Al_2(SO_4)_3$ and is commonly called filter alum. It is purchased as a liquid at a concentration of 48 percent alum in a water solution. This alum solution is then added to the incoming raw water at the rate of 18-24 mg/L. In everyday products, alums are used in baking powders, in deodorants, and to make pickles crisp. In water treatment, alum is used as a coagulant. A coagulant binds together very fine suspended particles into larger particles that can be removed by settling and filtration. In this way, objectionable color and turbidity (cloudiness), as well as the aluminum itself, are removed from the drinking water.

Lime

Hydrated lime is Calcium Hydroxide or $Ca(OH)_2$. It is an alkaline compound that is added to the water for pH adjustment. Since filter alum is an acidic salt, it lowers the pH of the water, and lime is used to neutralize this effect. Lime is added between the sedimentation and filtration processes at the rate of 10-20 mg/L.

Chlorine

Chlorine is added to water to disinfect it or improve its quality. Chlorine destroys pathogenic microorganisms, oxidizes undesirable elements such as iron and manganese and reduces some tastes and odors. After these needs have been met, some Chlorine remains (residual) in the water to protect it from further contamination until it reaches the customer's tap. Before the use of Chlorine, water-borne diseases such as typhoid fever were the source of devastating epidemics. Chlorine is the most important chemical added to water in terms of public health.

Three types of materials are commonly used as a source of Chlorine: gaseous Chlorine, Calcium Hypochlorite tablets, and Sodium Hypochlorite solution. Mainly because of safety concerns, Greensboro switched from Chlorine gas to Sodium Hypochlorite in 1999.

Sodium Hypochlorite, NaOCl, is a clear, greenish-yellow solution commonly used in bleaching. Household bleach usually contains about 5 percent available chlorine. For disinfection of drinking water, a 15 percent Sodium Hypochlorite solution is used to produce a 1-1.8 mg/L dosage of Chlorine in the finished water.

Polyphosphate

Calciquest is a liquid chemical solution containing a long chain polyphosphate polymer. Calciquest is designed to sequester iron, manganese and calcium to keep them in solution. It also provides a tough monomolecular film on the interior surface of water lines to prevent corrosion. As the water flows through the pipe, orthophosphate is deposited on the metal surface forming a thin film. As long as treatment is continued, the film is maintained and protects the metal from corrosive attack by water. This film does not build on itself to form a scale and thus does not obstruct the flow of water. Calciquest is added to the finished water at the rate of 1.0 mg/L.

Fluorid

Fluoridation is used to maintain fluoride concentrations in drinking water at levels known to reduce tooth decay in children. At optimum levels, fluoride can reduce the incidence of tooth decay among children by 65 percent. The amount of fluoride ingested depends on the amount of water consumption which usually depends on the temperature in a region. The State has recommended control limits of 0.7 to 1.2 mg/L fluoride. Greensboro has established 1.0 mg/L as the optimum fluoride level for this area. Fluoride is received as a 25 percent solution of hydrofluosilicic acid (H_2SiF_6). This solution is fed as a final treatment step before the water enters the distribution system.

Vocabulary

potassium permanganate	高锰酸钾	polyphosphate	聚磷酸盐
trihalomethane	三氯甲烷	fluoridation	氟化反应
deodorant	除臭剂,除臭的	hydrofluosilicic acid	氟硅酸
hypochlorite	次氯酸盐		

Unit 12 Sedimentation

Sedimentation or clarification is the removal of particulate matter, chemical floc, and precipitates from suspension through gravity settling. This makes water clarification a vitally important step in the treatment of surface waters for potable supply and in most cases the main factor in determining the overall cost of treatment. Poor design of the sedimentation basin will result in reduced treatment efficiency that may subsequently upset other operations.

Surface water containing high turbidity may require sedimentation prior to chemical treatment. Presedimentation basins are constructed in excavated ground or out of steel or concrete. Steel and concrete tanks are often equipped with a continuous mechanical sludge removal apparatus. The minimum recommended detention time for presedimentation is 3 h, although at many times of the year, this may not be adequate to remove fine suspensions. Chemical feed equipment may be provided ahead of presedimentation to provide prechlorination or partial coagulation for periods when water is too turbid to clarify by plain sedimentation.

Settling basins are usually provided for chemical coagulation or softening. These basins may be constructed of concrete or steel in a wide variety of shapes and flow mechanisms. To minimize the effects of short-circuiting and turbulent flow, care is taken in the effective hydraulic design of inlet and outlet structures in all tanks. Inlet structures are expected to: 1) uniformly distribute flow over the cross section of the settling zone; 2) initiate parallel or radial flow; 3) minimize large-scale turbulence; and 4) preclude excessive velocities near the sludge zone. Flow through a sedimentation basin enters at the top of the basin. In some circular, the flow may enter a central flocculating chamber of the basin, effluent flows vertically out over perimeter weirs. In rectangular tanks water flows horizontally to an end weir for discharge. An efficiently designed vertical tank is more stable than a horizontal one. The rating of a vertical tank as related to the bulk settling velocity is usually 1 to 3 m/h.

Recent developments have been made with the implementation of tube settling in the design of horizontal-flow settling tanks, which are said to overcome some of the previous difficulties and enable higher rates of sedimentation. Upflow solids contact clarifiers, and horizontal-flow basins may increase their volumetric capacity 50 to 150 percent or more with the introduction of inclined tubes into their basins. Similar or improved effluent quality is observed in these modified process. Effluent quality for any sedimentation process is dependent upon previous water quality, coagulants added, mixing, flocculation, and filtration. With the exception of presedimentation, sedimentation processes are usually preceded by coagulation and followed by filtration. Therefore, typical removals of regulated contaminants by sedimentation and filtration are reported according to coagulants used in conjunction with these processes for removal of a specific contaminant.

Basically, the theory of sedimentation is the theory of the effect of gravity on particles suspended in a liquid of lesser density. Under the influent of gravity, any particle having a density greater than 1.0 will settle in water at an accelerating velocity until the resistance of the liquid equals the effective weight of the particle. Thereafter, the settling velocity will be essentially constant and will depend upon the size, shape, and density of the particle, as well as the density and viscosity of the water. For most theoretical and practical computations of settling velocities in sedimentation basins, the shape of the particles is assumed to be spherical. Settling velocities of particles of other shapes can be analyzed in relation to spheres.

Number of Basins

Probably the most important considerations in the selection of the number of sedimentation basins are: 1) the effect upon the production of water if one basin is removed from service; and 2) the largest size which can be expected to produce satisfactory results. Plant layout and site conditions also may influence the number of basins. For any supply which requires coagulation and filtration for the production of safe water, a minimum of two basins should be provided. The greater the number available, the less will be the effect upon velocity and detention period in the remaining basins should one be out of service for cleaning, repairs, or any other reason.

Size and Shape of Basins

The most common forms of sedimentation basins are circular, square, or rectangular. The selection of the particular form or shape for a given plant depends upon area available, conformity with adjacent structures, and the theories and experience of the design engineer. The majority of sedimentation basins now in use are rectangular in shape and of reinforced concrete construction. Circular basins of either concrete or steel have been used widely for both coagulation and softening.

Most sedimentation basins are provided with sloping bottoms to facilitate the removal of deposited sludge. Although the optimum depth of a basins is a controversial matter, a review of the design of the larger rapid-sand filtration plants indicates a range of 2.43 to 4.88 m in depth. Multiple-story basins, in which the water travels horizontally along one level of a basin and then passes vertically (preferably upward) to another level of the basin for more horizontal flow, have been used successfully to compress a large amount of basin surface (bottom) area into a small area of the plant site. In this way, low values of the surface-loading parameter can be attained, producing high removal efficiencies. Since long, narrow tanks tend to have better hydraulic stability characteristics (less tendency to short-circuit) than short, wide ones, some designers place the several stories or levels in series rather than parallel. Although the same value for the surface-loading parameter can be attained with either flow arrangement, the parallel arrangement avoids the upsets by series arrangements in the area where flow direction reverses.

A recent innovation, the "tube" settler, uses parallel flow upward though tubes about 5 cm square inclined at an angle of 60° from the horizontal to attain extremely high surface areas. The tubes are short, 0.6 to 1.2 m, and the detention time for water in the "setter" is between 3 and 6 min. The 60° angle inclination permits accumulated sludge to slide down the tube wall to a sludge-collector area. An alternate design uses 2.5 cm hexagonal tubes inclined at 5° from the horizontal. In this case, the tube settler requires separate provision for sludge removal, commonly by backwashing with filter washwater. The steeple inclined tubes are constructed of plastic in modules which can be readily installed in existing sedimentation basins.

Vocabulary

gravity	重力	inclined tube	斜管
presedimentation	预沉淀	conformity with	与…一致
prechlorination	预氯化	sloping	倾斜的,有坡度的
short-circuiting	短流	surface-loading	表面负荷
preclude	排除	hexagonal	六角形的
rectangular	矩形	module	模数,模块

Reading Material A

Groundwater Quality

Like surface water, groundwater is vulnerable to contamination from a variety of sources. In Texas, all nine major aquifers and 20 minor aquifers have experienced some form of contamination. These contamination problems stem partly from land-based development and industry and partly from over-pumping, which causes infiltration of saline waters (see Fig. 12.1).

Despite these problems, state laws do not protect groundwater to the same extent as surface water. There are, for example, no groundwater quality standards to parallel those for surface water. The state does have standards in place for groundwater utilized for drinking water. In addition, monitoring of groundwater is divided among a myriad of state and local agencies.

The legislature has also taken steps to protect groundwater. In 1989 the legislature created the Texas Groundwater Protection Committee as an interagency committee to coordinate all state agency actions for the protection of groundwater quality. In 1991 the state created a $ 10-million oil-field cleanup fund to plug abandoned oil wells and pits. To protect endangered species and the sustainability of the Edwards Aquifer in San Antonio, the legislature in 1993 created the Edwards Aquifer Authority, which in turn has enacted a management plan and established a permit system for groundwater withdrawals to ensure adequate spring flows at the Comal and San Marcos springs. The law gives the Edwards Aquifer Authority the power to set limits on pumping.

In 1997 the legislature passed SB 1, commonly referred to as the "Water Bill". While the intent

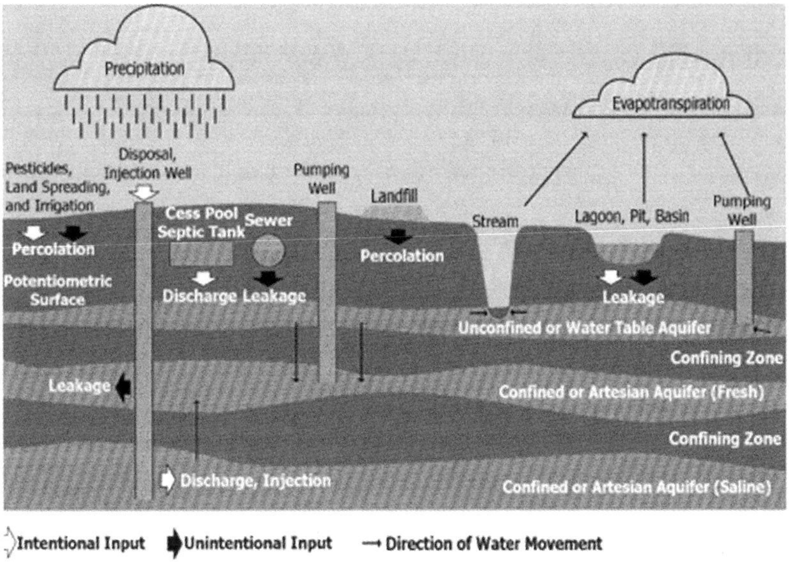

Fig.12.1 Typical routes of groundwater contamination

behind the bill is to ensure sufficient water quantity through a regional planning effort, drought management, conservation plans, and emergency authorizations of water transfers during droughts, the legislation also deals with water quality. Groundwater district management plans and regional water plans must address water quality. In addition, the bill authorizes the TWDB to administer the Federal Safe Drinking Water Revolving Fund to provide low-interest loans to small communities for drinking water and wastewater treatment. In 2001, under "Senate Bill 2", the Texas legislature clarified the authority of groundwater districts to regulate water wells as well as additional types of wells, and gave them the power to purchase groundwater rights for conservation, while also strengthening the enforcement powers of districts. These measures—among others—could allow districts to better protect water quality.

Clean groundwater is needed for more than drinking purposes. Agriculture depends heavily on groundwater for irrigation; in 1997 some 79 percent of all groundwater pumped was used for crop irrigation. Poor or contaminated groundwater could jeopardize crops and threaten the health of livestock. Clean groundwater is also essential to clean surface water. Groundwater is connected to surface water in the hydrological cycle, and some aquifers actually feed area springs and rivers. For example, the Edwards Aquifer is the major source for Central Texas rivers through the Comal and San Marcos springs. Poor quality water—or a lack of water—harms the springs.

Sometimes contamination occurs naturally. Saline water from deeper aquifers may reach aquifers that provide water for humans. Some groundwater may have naturally high background levels of nitrates, metals, iron, sulfate, or chloride, all of which can give water an odd odor, color, or taste.

Groundwater contamination has become a major public policy concern in recent decades. Human activity on virtually any piece of land in the state has the potential to affect groundwater quality, since 76 percent of the state's surface area of 267,277 square miles lies above major and minor aquifers. A variety of potential threats to groundwater are the result of human activities.

It is difficult to ascertain how much groundwater contamination has resulted from human activities. Under a 1991 state law, the legislature created the Texas Groundwater Protection Committee, which is formed by nine state agencies. As part of its duties, the committee tracks instances of groundwater contamination. The committee, utilizing data from all state agencies, reported 7,459 instances of groundwater contamination between 1989 and 1997 that had yet to be cleaned up. These figures represent only the cases reported and confirmed. In addition to these instances of groundwater contamination, action was completed on a total of 2,637 cases of groundwater contamination between 1989 and 1997.

Some of the leading sources of groundwater contamination in Texas include: 1) Underground and above-ground petroleum storage tanks; 2) Mining activities; 3) Various types of underground injection wells; 4) Water, oil and gas wells; 5) Existing and abandoned municipal and industrial waste facilities; 6) Agricultural activities.

Vocabulary

vulnerable	易于…的	jeopardize	危害
infiltration	渗滤	ascertain	确定
conservation	保持,保存		

Reading Material B

Pre-sedimentation and Sedimentation

Sedimentation, or clarification, is the process of solid particles from suspension by gravity. Suspended material may be clay or sludge. In drinking water treatment process, the common application of sedimentation is after chemical treatment to remove flocculated impurities and precipitate. In wastewater processing, sedimentation is used to reduce suspended solids.

Sedimentation is accomplished by decreasing the velocity of the water being treated to a point below which the particles will no longer remain in suspension. When the velocity no longer supports the transport of the particles, gravity will remove them from the flow.

The two principal applications of sedimentation in water treatment are plain sedimentation and sedimentation of coagulated and flocculated waters. Plain sedimentation is used to remove solids that are present in surface waters, and that settle without chemical treatment, such as gravel, sand, silt, and so on. It is used as a preliminary process to reduce the sediment loads in the remainder of the treatment plant, and is referred to as pre-sedimentation. Sedimentation also is used downstream of the coagulation and flocculation processes to remove solids that have been rendered more settleable by these process. Chemical coagulation may be geared toward removal of turbidity, color, or hardness.

Pre-sedimentation

The purpose of pre-sedimentation is to reduce the load of sand, silt, turbidity, bacteria, or other substances applied to subsequent treatment processes so that the subsequent processes may function more efficiently. When the raw water has exceptionally high concentrations of these substances, good removals are often obtained by plain settling and without the use of chemicals. Waters containing gravel, sand, silt, or turbidity in excess of 1,000 NTU may require pretreatment.

Pre-sedimentation basins should have hopper bottoms and/or be equipped with continuous mechanical sludge removal apparatus especially selected or designed to remove heavy silt or sand. Sludge is not removed continuously, and allowance must be made for sludge accumulation between cleanings. In manually cleaned basins, settled matter is often allowed to accumulate until it tends to impair the settled-water quality, at which time the sludge is flushed out.

Sedimentation basins that are not equipped for mechanical removal of sludge should have sloping bottoms, so that they can be rapidly drained, allowing most of the sediment to flow out with the water. Because the bulk of the material settles near the inlet end, the slope should be greatest at this point. In some plants, sluice gates are arranged to deliver raw water to the sedimentation basin to flush out the sludge. The balance is generally washed out with a fire hose. At least two basins are needed so one can remain in service while the other is cleaned.

The time between cleanings varies from a few weeks, in plants that have short periods of settling and handle very turbid water, to a year or more, where the basin capacity is large and the water is not very turbid.

Because the particles to be removed are more readily settleable than chemical flocs, the detention time may be shorter and the surface overflow and weir rates may be greater for pre-sedimentation basins than for primary or secondary settling basins. Detention times of at least 2 hours, maximum overflow rates of 5.9 m/h, maximum weir rates of 32.2 m/h, and minimum water depths of 0.91 m have been recommended as design standards for pre-sedimentation basins. When mechanical sludge removal is not provided, detention times of 2 to 3 days are often used to allow for sludge storage. Facilities for chlorination of the pre-sedimentation basin influent are provided in many cases.

Pre-sedimentation in itself can provide reductions in coliform organisms. Reported removals are:

1. Ninety percent or more for heavy coliform loadings (20,000 to 60,000 Most Probable Number (MPN) per 100 ml) and long pre-sedimentation periods (5 to 10 days).

2. About 20 percent for light coliform loadings (5,000 to 20,000 MPN per 100 ml) and shorter detention times (3 to 7 hours).

The relatively long storage periods provided by pre-sedimentation are also valuable in providing time for the natural die-away of virus that are in an unfriendly environment.

Sedimentation Following Coagulation and Flocculation

A common practice is to follow flocculation with sedimentation in order to reduce the load of solids applied to filter. A major portion of the solids can be removed from the water by gravity settling following coagulation and flocculation. As discussed in Unit 13, development of coarse-to-fine filters, through the use of mixed-media or dual-media filters, means that much heavier loads of suspended solids can be filtered at reasonable head losses than could be handled in the past using single-media, fine-to-coarse filters. This makes it possible to consider direct filtration (without settling following flocculation) in an increasing number of situations. However, there remains a need for sedimentation in waters that are too turbid for direct filtration or that require other chemical (i.e., softening) prior to filtration.

Vocabulary

accomplish	完成	detention time	停留时间
preliminary	初步的	coliform	大肠菌
settleable	可沉的	mixed-media	混合介质
concentration	浓度	dual-media	双层介质
accumulation	堆积	direct filtration	直接过滤

Unit 13 Filtration and Filter Types

The term "filtration" has different connotations. Outside the waterworks profession, even in other technical disciplines, filtration is commonly thought of as a mechanical straining process. It may also have this same basic meaning in waterworks practice as applied to the passage of water through a very thin layer of porous material deposited by flow on a support material. This type of filter has a few rather specialized applications to water treatment, as described later. Most frequently in waterworks parlance, filtration refers to the use of a relatively deep (0.46 to 0.91 m) granular bed to remove particulate impurities from water. In contrast to mechanical strainers that remove only part of the coarse suspended solids, the filters used in water purification can remove all suspended solids, including virtually all colloidal particles.

Over the years the meaning of the word "filtration" as used by the waterworks industry has changed. Improved filters have been developed, and the nature of the physical and chemical processes involved in filtration has become better understood. New and improved filters are distinguished from older conventional types because they can incorporate: 1) coarse-to-fine in depth filtration; 2) the application to the filter influent of a polymer, alum, or activated silica as a filter aid; 3) continuous monitoring of filter effluent turbidity; and 4) pilot filter control of coagulant dosage. Use of the general term "filters" overlooks important distinctions in functions and efficiency. In discussing filtration then, care must be exercised to specify the details of the particular filter under consideration.

Water filtration is a physical-chemical process for separating suspended and colloidal impurities from water by passage through a bed of granular material. There are two separate steps: 1) transport, in which suspended particles are transported to the immediate vicinity of the solid filter media; and 2) attachment, in which particles become attached to the filter media surface or to another particle previously retained in the filter. The transport step is primarily a physical process, while the attachment step is very much influenced by chemical and physical-chemical variables.

There are several ways to classify filters. They can be described according to the direction of flow through the bed, that is, downflow, upflow, radial flow, horizontal flow, fine-to-coarse, or coarse-to-fine. They may be classed according to the type of filter media used, such as sand, coal (or anthracite), coal-sand, multilayered mixed-media, or diatomaceous earth. Filters are also classed by flow rate. Slow sand filters operate at rates of 0.12 to 0.32 m/h, rapid sand filters operate at rates of 2.44 to 4.88 m/h, and high-rate filters operate at rates of 7.32 to 36.6 m/h. Filters may also be classified by the type of system used to control the flow rate through the filter, such as constant rate, declining rate, constant level, equal loading, and constant pressure. Constant rate filtration is the most popular control system in the United States.

Another characteristic is pressure or gravity flow. Gravity filter units are usually built with an open top and constructed of concrete or steel, while pressure filters are ordinarily fabricated from steel in the form of a cylindrical tank. The available head for gravity flow usually is limited to about 2.44 to 3.66 m, while it may be as high as 1,033 kPa for pressure filters. Because pressure filters have a closed top, it is not easy to inspect the filter media. Further, it is possible to disturb the media in a pressure filter by sudden changes in pressure. These two factors have tended to limit municipal applications of pressure filters to treatment of relatively unpolluted waters, such as the removal of hardness, iron, or manganese from well waters of good bacterial quality. The susceptibility to bed upset and the inability to see the media in pressure filters have been compensated for, to some extent, by the use of quick-opening manholes and by the recent development and application of recording turbidimeters for continuous monitoring of the filter effluent turbidity. The introduction of a 76.2-mm layer of coarse (1 mm) high-density (specific gravity 4.2) garnet or eliminated between the fine media and the gravel supporting bed has virtually eliminated the problem of gravel upsets, which is another of the concerns about the use of pressure filters for production of potable water.

In the examination of filter sand, the terms "effective size" and "uniformity coefficient" are used. The effective size is the size of the grain, in millimeters, such that 10 percent of the particles by weight are smaller. This 10 percent by weight fraction corresponds closely in size to the median size by count, as determined by a size frequency distribution of the total number of particles in a sample. The effective size is a good indicator of the hydraulic characteristics of a sand within certain limits. These limits are usually defined by means of the uniformity coefficient, which is arbitrarily taken as the ratio of the grain size with 60 percent smaller than itself to the size with 10 percent smaller than itself (effective size). This ratio thus covers the range in size of half the sand.

Vocabulary

connotation	内涵	attachment	吸附，黏附
porous	多孔的	radial flow	辐流
strainer	滤网	upflow	上向流
depth filtration	深层过滤	diatomaceous	硅藻的，硅藻土的，含硅藻的
activated silica	活化硅酸		
transport	转输	turbidimeter	浊度计

Reading Material A

Direct Filtration

As noted above, the ability of mixed-media filters to tolerate higher applied turbidities has resulted in several applications where coagulated water is filtered directly without sedimentation. Fig. 13.1 presents approaches to direct filtration now in use.

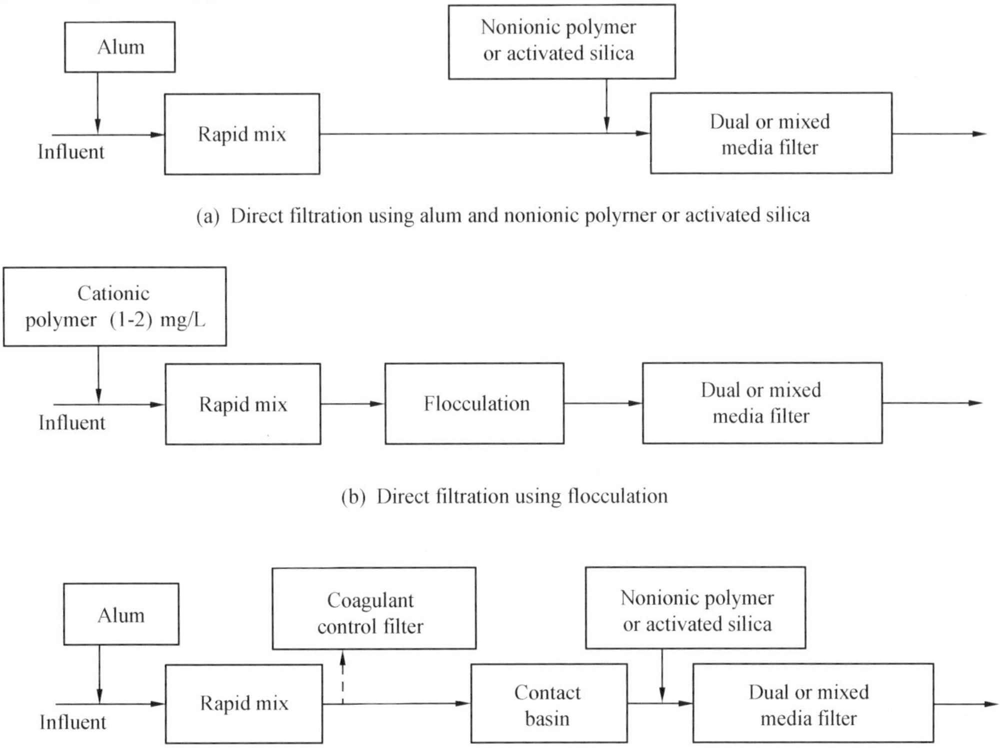

(a) Direct filtration using alum and nonionic polyrner or activated silica

(b) Direct filtration using flocculation

(c) Direct filtration using a contact basin

Fig. 13.1 Approaches to direct filtration

The direct filtration arrangement shown in Fig. 13.1(a) consists of the addition of alum to rapid mix influent followed by the application of a nonionic polymer or activated silica to the influent of dual- or mixed-media filters. An alternative would replace the alum and nonionic polymer with the use of a cationic polymer in the raw water entering the rapid mix. In some cases, preozonation has been reported to enhance the removal of turbidity in the direct filtration process. Potential mechanisms are micro-flocculation and the production of more polar compounds. Fig. 13.1(b) is a direct filtration scheme utilizing a flocculation basin in which the chemical dosage alternates are the same as for Fig. 13.1(a).

The Fig. 13.1(c) flow sheet is a direct filtration arrangement with a 1-hour contact basin between the rapid mix and the filter. In this process, the purpose of the contact basin is not to provide for settling. There is no flocculation basin, and the contact basin is not equipped with a sludge collector. The purpose of the contact basin is to increase the reliability of the process by adding a lead time of 1 hour between turbidity readings showing the results of coagulation from the coagulant-control filter and the entry of water to the filters. This helps in keeping plant operations abreast of changes in raw water quality. The coagulation options are the same as for Fig. 13.1 (a).

Following the proper mixing of the coagulant with raw water, a number of complex reactions take place with colloidal turbidity and color. These coagulation reactions take place in less than 1 s. The rapid mixing process for direct filtration usually does not differ from that for conventional plants. A hydraulic jump in a Parshall flume may be used; field experience in direct filtration plants has been good with this type of mixing device. Some engineers extend the time for mechanical rapid mixing in direct filtration plants up to as much as 5 minutes, which is longer than that used in most conventional plants.

At this point in the process, the particles formed are very small, and the colloids are destabilized. When the destabilized particles collide, they stick together, with the rate of agglomeration of these microscopic destabilized particles to form visible floc depending principally upon the number of opportunities for contact they are afforded. In a still body of water, agglomeration takes place at a slow, almost imperceptible rate; the rate can be increased by agitation or stirring of the water. In a well-designed flocculator, agglomeration of all particles might be completed in times varying from 5 to 45 minutes, when enough collisions will have occurred for the floc particles to become large enough to settle rapidly.

If settling is omitted from the plant flow sheet, as in a direct filtration plant, and if a properly designed rapid mix is provided, then usually there is no reason to include flocculation in the direct filtration process. Rather than spending money on flocculation, it may be better to improve the rapid mixing, provided finer filter media, or increase the depth of the fine filter media. The water containing the destabilized particles can be taken directly from the rapid mix basin to a granular filter where contact flocculation takes place as part of the filtration process. The flocculation is greatly accelerated because of the tremendous number of opportunities for contact afford in the passage of the water through the granular bed. The floc particles become attached or absorbed to the surface of the filter grains. The smaller the filter grains, the greater are the opportunities for contact, and the more rapid is the removal of particulate matter. Small filter media also have a greater surface area per unit volume, which provides more area for attachment of floc particles to the filter grains than is available with larger grains.

This contact and the surface attachment or adsorption of particles to filter grains account for particulate removal beyond that of any simple mechanical straining action of the fine media. The pores of the filter gradually fill with floc as particles are sheared off the surfaces of grains. As a filter run progresses, the upper pores of the filter cannot retain any more floc, and the headloss through the bed increases to the point where the filter must be cleaned by backwashing, or if the floc strength is inadequate because of an underfeed of polymer, floc particles may appear in the effluent, requiring filter backwashing.

For many raw waters, direct filtration water of a quality equal to that obtained by flocculation, settling, and filtration. The limitation of direct filtration is the inability to handle high concentrations of suspended solids. At some point, the amount of suspended solids will be too high for reasonable filter runs, and settling before filtration will be necessary.

The possibilities of applying direct filtration to municipal plants are good if one of the following conditions:

1. The raw water turbidity and color are each less than 25 NTU.
2. The color is low, and the maximum turbidity does not exceed 200 NTU.
3. The turbidity is low, and the maximum color does not exceed 100 NTU.

The presence of paper fiber or of diatoms in excess of 1,000 areal standard units per milliliter (asu/ml) requires that settling be included in the treatment process chain. Diatom levels in excess of 200 asu/ml may require the use of special coarse coal on the top of the bed in order to extend filter runs.

The suitability of a raw water for direct filtration cannot be determined from numerical values alone; such values only provide a preliminary indication. Pilot plant tests must be performed in each case to find out whether or not direct filtration will provide satisfactory treatment under the prevailing local circumstances of raw water quality.

The chief advantage of direct filtration is the potential for capital cost savings from the elimination of sludge-collecting equipment, settling basin structures, flocculation equipment, and flocculation-basin structures. This cost reduction may make possible the provision of needed filtration for some communities that could not otherwise afford it. Operation and maintenance costs are reduced because there is less equipment to operate and maintain.

With direct filtration there may also be a savings of 10 to 30 percent in chemical costs because generally less alum is required to produce a filterable floc than to produce a settleable floc. The costs for polymer may be greater than in conventional plants, but these higher costs are more than offset by the lower costs for coagulants.

Direct filtration produces less sludge than conventional treatment, and a denser sludge. The collection of waste solids is simplified. Waste solids are all contained in a single stream, the waste filter-backwash water.

Pilot plant tests and plant-operating experience show that high-quality filtrate can be produced in direct filtration at filter rates of 12.2 to 36.6 m/h. A usual design rate is 12.2 m/h. This provides for flexibility of operation and affords a margin of safety against variations in raw water quality. Filter influent and effluent piping should be designed for flows of 24.4 m/h.

The design considerations presented below apply to filters used in direct filtration systems as well as to those used downstream of clarifiers.

Vocabulary

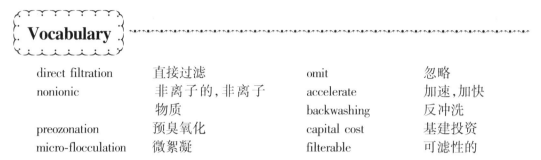

direct filtration	直接过滤	omit	忽略
nonionic	非离子的,非离子物质	accelerate	加速,加快
		backwashing	反冲洗
preozonation	预臭氧化	capital cost	基建投资
micro-flocculation	微絮凝	filterable	可滤性的

Reading Material B

Filter Backwashing

During the service cycle of filter operation, particulate matter removed from the applied water accumulates on the surface of the grains of fine media and in the pore spaces between grains. With continued operation of a filter, the materials removed from the water and stored within the bed reduce the porosity of the bed. This has two effects on filter operation: it increases the headloss through the filter, and it increases the shearing stresses on the accumulated floc. Eventually the total hydraulic headloss may approach or equal the head necessary to provide the desired flow rate through the filter, or there may be a leakage or breakthrough of floc particles into the filter effluent. Just before either if these outcomes can occur, the filter should be removed from service for cleaning. In the old slow-sand filters the arrangement of sand particles is fine to coarse in the direction of filtration (down); most of the impurities removed from the water collect on the top surface of the bed, which can be cleaned by mechanical scraping and removal of about 12.7 mm of sand and floc. In rapid sand filters, there is somewhat deeper penetration of particles into the bed because of the coarser media used and the higher flow rates employed. However, most of the materials are stored in the top few inches of a rapid sand filter bed. In dual-media and mixed-media beds, floc is stored throughout the bed depth to within a few inches of the bottom of the fine media.

Rapid sand, dual-media, and mixed-media filters are cleaned by hydraulic backwashing (upflow) with potable water. Thorough cleaning of the bed makes it advisable in the case of single-medium filters and mandatory in the case of dual-or mixed-media filters to use auxiliary scour or so-called surface wash devices before or during the backwash cycle. Backwash flow rates of 36.6 to 48.8 m/h should be provided. A 20 to 50 percent expansion of the filter bed is usually adequate to suspend the bottom grains. The optimum rate of washwater application is a direct function of water temperature, as expansion of the bed varies inversely with viscosity of the washwater. For example, a backwash rate of 43.9 m/h at 20 ℃ equates to 38.3 m/h at 5 ℃ and 48.8 m/h at 35 ℃. The time required for complete washing varies from 3 to 15 minutes.

Following the washing process, water should be filtered to waste until the turbidity drops to an acceptable value. Filter-to-waste outlets should be through an air-gap-to-waste drain, which may require from 2 to 20 minutes, depending on pretreatment and type of filter. This practice was discontinued for many years, but modern recording turbidimeters have shown that this operation is valuable in the production of a high-quality water. Operating the washed filter at a slow rate at the start of a filter run may accomplish the same purpose. A recording turbidimeter for continues monitoring of the effluent from each individual filter unit is of great value in controlling this operation at the start of a run, as well as in predicting or detecting filter breakthrough at the end of a run.

As backwashing begins, the sand grains do not move apart quickly and uniformly throughout the bed. Time is required for the sand to equilibrate at its expanded spacing in the upward flow of washwater. If the backwash is turned on suddenly, it lifts the sand bed bodily above the gravel layer, forming an open space between the sand and gravel. The sand bed then breaks at one or more points, causing sand boils and subsequent upsetting of the supporting gravel layers, so that the gravel section must be rebuilt. It is essential that the backwash valve open slowly.

The time from start to full backwash flow should be at least 30 seconds and perhaps longer, and should be restricted by devices built into the plant. This is frequently done by means of an automatically regulated master wash valve, controlled hydraulically or electrically and designed so that it cannot open too fast. Alternatively, a speed controller could be installed on the operator of each washwater valve.

Filters can be seriously damaged by slugs of air introduced during filter backwashing. The supporting gravel can be overturned and mixed with the fine media, which requires removal and replacement of all media for proper repair. Air can be unintentionally introduced to the bottom of the filter in a number of ways. If a vertical pump is user for the backwash supply, air may collect in the vertical pump column between backwashings. The air can be eliminated without harm by starting the pump against a closed discharge valve. And bleeding the air out from behind the valve through a pressure air release valve. The pressure air release valve must have sufficient capacity to discharge the accumulated air in a few seconds.

Also, air or dissolved oxygen, released from the water on standing and warming in the washwater supply piping, may accumulated at high points in the piping and be swept into the filter underdrains by the inrushing washwater. This can be avoided by placing a pressure air release valve at the high point in the line, and providing a 12.7-mm pressure water connection to the washwater supply header to keep the line full of water and to expel the air.

The entry for washwater into the filter bottom must be designed to dissipate the velocity head of the washwater in such a manner that uniform distribution of washwater is obtained. Lack of attention to this important design factor has often led to difficult and expensive alterations and corrective repairs to filters.

Horizontal centrifugal pumps may be used to supply water from the treated water storage reservoir for filter backwashing, provided they are located where positive suction head is available, or are provided with an adequate priming system. Otherwise, a vertical pump unit may be suspended in the clearwell. The pumps should be installed with an adequate air release valve, a nonslam check valve, and a throttling valve in the discharge. Single washwater supply pumps are often installed, but consideration should be given to provisions for standby service, including multiple pump units and /or standby generators.

Washwater may also be supplied by gravity flow from a storage tank located above the top of the filter boxes. Washwater supply tanks usually have a minimum capacity equal to a 7-minutes for one filter unit, but may be larger. The bottom of the tank must be high enough above the filter wash troughs to supply water at the rate required for backwashing as determined by a hydraulic analysis of the washwater system. This distance is usually at least 3.05 m, but more often is 7.6 m or greater. Washwater tanks should be equipped with an overflow line, and a vent for release and admission of air above the high-water level. Washwater tanks are often filled by means of a small electric driven pump equipped with a high-water level cutoff. They may also be filled from the high-service pump discharge line through an altitude or float valve, but this is usually wasteful of electrical energy. Washwater tanks may be constructed of steel or concrete. Provisions should be made for shutdown of the tanks for maintenance and repair.

The washwater supply line must be equipped with provisions for accurately (within percent) measuring and regulating the rate of washwater flow. The rate of flow indicator should be visible to the point at which the rate of flow is regulated.

The use of a high pressure, above 103.4 kPa, source of filter backwash water through a pressure-reducing valve is not advised. Numerous failures of systems using pressure-reducing valves have so thoroughly upset and mixed the supporting gravel and fine media that these materials have had to be completely removed from the filter and replace with new media.

In some cases, pollution control regulations prohibit the once common practice of discharging backwash wastewaters directly to a stream. In these cases, the backwash wastewater must be reprocessed. The rate of backwash flow, if it is returned directed to an upstream clarifier, is usually large enough in relation to the design flow through the clarifier to cause a hydraulic overload and upset of the clarifier. In this case, the back wash should be collected in a storage tank and recycled at a controlled rate. The volume of backwash wastewater is typically 2 to 5 percent of the plant throughout, and plant components must be sized to handle this recycled flow.

Vocabulary

porosity	多孔性，有孔性，孔隙度	hydraulic	水力的
headloss	水头损失	scour	擦洗
shearing stress	剪切力	eliminate	消除
breakthrough	穿透		

Unit 14 Disinfection

The treatment and distribution of water for safe use is one of the greatest achievements of the twentieth century. Before cities began routinely treating drinking water with chlorine (starting with Chicago and Jersey City in 1908), cholera, typhoid fever, and dysentery killed thousands of U. S. residents annually. Drinking water chlorination and filtration have helped to virtually eliminate these diseases in the U.S. and other developed countries.

Meeting the goal of clean, safe drinking water requires a multi-barrier approach that includes: protecting source water from contamination, appropriately treating raw water, and ensuring safe distribution of treated water to consumers' taps.

During the treatment process, chlorine is added to drinking water as elemental chlorine (chlorine gas), sodium hypochlorite solution or dry calcium hypochlorite. When applied to water, each of these forms "free chlorine," which destroys pathogenic (disease-causing) organisms.

Almost all U.S. systems that disinfect their water use some type of chlorine-based process, either alone or in combination with other disinfectants. In addition to controlling disease-causing organisms, chlorination offers a number of benefits including:

- Reduces many disagreeable tastes and odors;
- Eliminates slime bacteria, molds and algae that commonly grow in water supply reservoirs, on the walls of water mains and in storage tanks;
- Removes chemical compounds that have unpleasant tastes and hinder disinfection; and
- Helps remove iron and manganese from raw water.

As importantly, only chlorine-based chemicals provide "residual disinfectant" levels that prevent microbial re-growth and help protect treated water throughout the distribution system.

Concentration and Contact Time

In an attempt to establish more structured operating criteria for water treatment disinfection, the CXT concept came into use in 1980. Based on the work of several researchers, CXT values (final free chlorine concentration (mg/L) multiplied by minimum contact time (minutes)), offer water operators guidance in computing an effective combination of chlorine concentration and chlorine contact time required to achieve disinfection of water at a given temperature. The CXT formula demonstrates that if an operator chooses to decrease the chlorine concentration, the required contact time must be lengthened. Similarly, as higher strength chlorine solutions are used, contact times may be reduced (Connell, 1996).

The Challenge of Disinfection Byproducts

While protecting against microbial contamination is the top priority, water systems must also control disinfection byproducts (DBPs), chemical compounds formed unintentionally when chlorine and other disinfectants react with natural organic matter in water. In the early 1970s, EPA scientists first determined that drinking water chlorination could form a group of byproducts known as trihalomethanes (THMs), including chloroform. EPA set the first regulatory limits for THMs in 1979. While the available evidence does not prove that DBPs in drinking water cause adverse health effects in humans, high levels of these chemicals are certainly undesirable. Cost-effective methods to reduce DBP formation are available and should be adopted where possible. However, a report by the International Program on Chemical Safety (IPCS 2000) strongly cautions:

The health risks from these byproducts at the levels at which they occur in drinking water are extremely small in comparison with the risks associated with inadequate disinfection. Thus, it is important that disinfection not be compromised in attempting to control such byproducts.

Recent EPA regulations have further limited THMs and other DBPs in drinking water. Most water systems are meeting these new standards by controlling the amount of natural organic material prior.

Controlling Disinfection Byproducts

Treatment techniques are available that provide water suppliers the opportunity to maximize potable water safety and quality while minimizing the risk of DBP risks. Generally, the best approach to reduce DBP formation is to remove natural organic matter precursors prior to disinfection. EPA has published a guidance document for water system operators entitled, Controlling Disinfection Byproducts and Microbial Contaminants in Drinking Water (EPA, 2001).

The EPA guidance discusses three processes to effectively remove natural organic matter prior to disinfection:

1. Coagulation and Clarification: Most treatment plants optimize their coagulation process for turbidity (particle) removal. However, coagulation processes can also be optimized for natural organic matter removal with higher doses of inorganic coagulants (such as alum or iron salts), and optimization of pH.

2. Absorption: Activated carbon can be used to absorb soluble organics that react with disinfectants to form byproducts.

3. Membrane Technology: Membranes, used historically to desalinate brackish waters, have also demonstrated excellent removal of natural organic matter. Membrane processes use hydraulic pressure to force water through a semi-permeable membrane that rejects most contaminants. Variations of this technology include reverse osmosis (RO), nanofilitration (low pressure RO), and microfiltration (comparable to conventional sand filtration).

Other conventional methods of reducing DBP formation include changing the point of chlorination and using chloramines for residual disinfection. EPA predicts that most water systems will be able to achieve compliance with new DBP regulations through the use of one or more of these relatively low cost methods (EPA, 1998).

Water system managers may also consider switching from chlorine to alternative disinfectants to reduce formation of THMs and HAAs. However, all chemical disinfectants form some DBPs. Much less is known about the byproducts of these alternatives than is known about chlorination byproducts. Furthermore, each disinfection method has other distinct advantages and disadvantages.

The Future of Chlorine Disinfection

Despite a range of new challenges, drinking water chlorination will remain a cornerstone of waterborne disease prevention. Chlorine's wide array of benefits cannot be provided by any other single disinfectant. While alternative disinfectants (including chlorine dioxide, ozone, and ultraviolet radiation) are available, all disinfection methods have unique benefits, limitations, and costs. Water system managers must consider these factors, and design a disinfection approach to match each system's characteristics and source water quality.

In addition, world leaders increasingly recognize safe drinking water as a critical building block of sustainable development. Chlorination can provide cost-effective disinfection for remote rural villages and large cities alike, helping to bring safe water to those in need.

Vocabulary

chlorination	氯化处理	membrane	膜
sodium hypochlorite	次氯酸钠	reverse osmosis	反渗透
pathogenic	致病的	nanofilitration	纳滤
disinfection byproducts	消毒副产物	microfiltration	微滤
trihalomethanes	三氯甲烷		

Reading Material A

Comparing Alternative Disinfection Methods

Up until the late 1970s, chlorine was virtually the only disinfectant used to treat drinking water. Chlorine was considered an almost ideal disinfectant, based on its proven characteristics:

- Effective against most known pathogens;
- Provides a residual to prevent microbial re-growth and protect treated water throughout the distribution system;
- Suitable for a broad range of water quality conditions;
- Easily monitored and controlled;
- Reasonable cost.

More recently, drinking water providers have faced an array of new challenges, including:

- Treating resistant pathogens such as Giardia and Cryptosporidium;

- Minimizing disinfection byproducts;
- New environmental and safety regulations;
- Strengthening security at treatment facilities.

To meet these new challenges, water system managers must design unique disinfection approaches to match each system's characteristics and source water quality. While chlorination remains the most commonly used disinfection method by far, water systems may use alternative disinfectants, including chloramines, chlorine dioxide, ozone, and ultraviolet radiation. No single disinfection method is right for all circumstances, and in fact, water systems may use a variety of methods to meet overall disinfection goals at the treatment plant, and to provide residual protection throughout the distribution system.

The sections below describe various disinfection technologies, and discuss the major advantages and limitations associated with each.

Chlorination

Chlorine is applied to water in one of three forms: elemental chlorine (chlorine gas), hypochlorite solution (bleach), or dry calcium hypochlorite. All three forms produce free chlorine in water.

Advantages

- Highly effective against most pathogens;
- Provides a residual to protect against recontamination and to reduce bio-film growth in the distribution system;
- Easily applied, controlled, and monitored;
- Strong oxidant meeting most preoxidation objectives;
- Operationally the most reliable;
- The most cost-effective disinfectant.

Limitations

- Byproduct formation (THMs, HAAs);
- Will oxidize bromide to bromine, forming brominated organic byproducts;
- Not effective against Cryptosporidium;
- Requires transport and storage of chemicals.

Elemental Chlorine

Elemental chlorine is the most commonly used form of chlorine. It is transported and stored as a liquefied gas under pressure. Water treatment facilities typically use chlorine in 100 and 150 L cylinders or one-ton containers. Some large systems use railroad tank cars or tanker trucks.

Advantages

- Lowest cost of chlorine forms;
- Unlimited shelf-life.

Limitations

● Hazardous gas requires special handling and operator training;

● Additional regulatory requirements, including EPA's Risk Management Program and the Occupational Safety and Health Administration's Process Safety Management Program.

Sodium Hypochlorite

Sodium Hypochlorite, or bleach, is produced by adding elemental chlorine to sodium hydroxide. Typically, hypochlorite solutions contain from 5 to 15 percent chlorine, and are shipped by truck in one-to 5,000-gallon containers.

Advantages

● Solution is less hazardous and easier to handle than elemental chlorine;

● Fewer training requirements and regulations than elemental chlorine.

Limitations

● Limited shelf-life;

● Potential to add inorganic byproducts (chlorate, chlorite and bromate) to water;

● Corrosive to some materials and more difficult to store than most solution chemicals;

● Higher chemical costs than elemental chlorine.

Chloramines

Chloramines are chemical compounds formed by combining a specific ratio of chlorine and ammonia in water. Because chloramines are relatively weak as a disinfectant, they are almost never used as a primary disinfectant. Chloramines provide a durable residual, and are often used as a secondary disinfectant for long distribution lines and where free chlorine demand is high. Chloramines may also be used instead of chlorine in order to reduce chlorinated byproduct formation and to remove some taste and odor problems.

Advantages

● Reduced formation of THMs, HAAs;

● Will not oxidize bromide to bromine forming brominated byproducts?

● More stable residual than free chlorine;

● Excellent secondary disinfectant, has been found to be better than free chlorine at controlling coliform bacteria and biofilm growth;

● Lower taste and odor than free chlorine.

Limitations

● Weak disinfectant and oxidant;

● Requires shipment and handling of ammonia or ammonia compounds as well as chlorinating chemicals;

● Ammonia is toxic to fish, and may pose problems for aquarium owners;

● Will cause problems for kidney dialysis if not removed from water.

Chlorine Dioxide

Chlorine dioxide (ClO_2) is generated on-site at water treatment facilities. In most generators sodium chlorite and elemental chlorine are mixed in solution, which almost instantaneously forms chlorine dioxide. Chlorine dioxide characteristics are quite different from chlorine. In solution it is a dissolved gas, which makes it largely unaffected by pH but volatile and relatively easily stripped from solution. Chlorine dioxide is also a strong disinfectant and a selective oxidant. While chlorine dioxide does produce a residual it is only rarely used for this purpose.

Advantages

- Effective against Cryptosporidium;
- Up to five times faster than chlorine at inactivating Giardia;
- Disinfection is only moderately affected by pH;
- Will not form chlorinated byproducts (THMs, HAAs);
- Does not oxidize bromide to bromine (can form bromate in sunlight);
- More effective than chlorine in treating some taste and odor problems;
- Selective oxidant used for manganese oxidation and targeting some chlorine resistant organics.

Limitations

- Inorganic byproduct formation (chlorite, chlorate);
- Highly volatile residuals;
- Requires on-site generation equipment and handling of chemicals (chlorine and sodium chlorite);
- Requires a high level of technical competence to operate and monitoring equipment, product and residuals;
- Occasionally poses unique odor and taste problems;
- High operating cost (chlorite chemical cost is high).

Ozone

Ozone (O_3) is generated on-site at water treatment facilities by passing dry oxygen or air through a system of high voltage electrodes. Ozone is one of the strongest oxidants and disinfectants available. Its high reactivity and low solubility, however, make it difficult to apply and control. Contact chambers are fully contained and non-absorbed ozone must be destroyed prior to release to avoid corrosive and toxic conditions. Ozone is more often applied for oxidation rather than disinfection purposes.

Advantages

- Strongest oxidant/disinfectant available;
- Produces no chlorinated THMs, HAAs;
- Effective against Cryptosporidium at higher concentrations;
- Used with Advanced Oxidation processes to oxidize refractory organic compounds.

Limitations

- Process operation and maintenance requires a high level of technical competence;
- Provides no protective residual;
- Forms brominated byproducts (bromate, brominated organics);
- Breaks down more complex organic matter; smaller compounds can enhance microbial re-

growth in distribution systems and increase DBP formation during secondary disinfection processes;

- Higher operating and capital costs than chlorination;
- Difficult to control and monitor particularly under variable load conditions.

Ultraviolet Radiation

Ultraviolet (UV) radiation, generated by mercury arc lamps, is a non-chemical disinfectant. When UV radiation penetrates the cell wall of an organism, it damages genetic material, and prevents the cell from reproducing. Although it has a limited track record in drinking water applications, UV has been shown to effectively inactivate many pathogens while forming limited disinfection byproducts.

Advantages

- Effective at inactivating most viruses, spores and cysts;
- No chemical generation, storage, or handling;
- Effective against Cryptosporidium;
- No known byproducts at levels of concern.

Limitations

- No residual protection;
- Low inactivation of some viruses (reoviruses and rotaviruses);
- Difficult to monitor efficacy;
- Irradiated organisms can sometimes repair and reverse the destructive effects of UV through a process known as photo-reactivation;
- May require additional treatment steps to maintain high-clarity water;
- Does not provide oxidation, or taste and odor control;
- High cost of adding backup/emergency capacity;
- Mercury lamps may pose a potable water and environmental toxicity risk.

Vocabulary

bio-film	生物膜	chloramines	氯氨
cost-effective	经济的,划算的	oxidant	氧化剂
bromide	溴化物	reovirus	呼吸道肠道病毒
bromine	溴	rotavirus	轮状病毒
cryptosporidium	隐孢子		

Reading Material B

Disinfection By-products

For centuries, people were largely unaware of the need to treat the water we drink. Many thought that the taste of the water determined its purity, not knowing that even the best tasting water could contain disease-causing organisms. Even the fact that disease could be spread through drinking water was not commonly known until the latter part of the 1800s.

With this knowledge came an awareness of the need to treat our water. Great Britain began disinfecting its drinking water early in the 20th century and saw a sharp decline in typhoid deaths. Shortly after, disinfection was introduced into the United States, which resulted in the virtual elimination of waterborne diseases such as cholera, typhoid, dysentery, and hepatitis A. The goal of disinfection is to kill or render harmless microbiological organisms that cause disease and immediate illness.

The most common method of disinfection is through the addition of chlorine to drinking water supplies. Not only is chlorine effective against waterborne bacteria and viruses in the source water, it also provides residual protection to inhibit microbial growth after the treated water enters the distribution system. This means it continues working to keep the water safe as it travels from the treatment plant to the consumer's tap.

However, even though chlorine has been a literal lifesaver with regard to drinking water, it also has the potential to form by-products that are know to produce harmful health effects. Chlorine can combine with organic materials in the raw water to create a contaminant called trihalomethanes (THMs). Repeated exposure to elevated levels of THMs over a long period of time could increase a person's risk of cancer.

The formation of disinfection by-products is a greater concern for water systems that use surface water, such as rivers, lakes, and streams, as their source. Surface water sources are more likely to contain the organic materials that combine with chlorine to form THMs.

Surface water systems serving a population greater than 10,000 must regularly test their treated water to determine if THMs are present. If the THMs exceed the limits set by the U.S. Environmental Protection Agency (EPA), the water system must take action to correct the problem. The corrective actions include notifying all residents served by the water system. By 2003, all public water systems that provide disinfection will be required to test for THMs and meet the limits set by EPA.

Since the mid-1970s, when the threat posed by disinfection by-products became known, water utilities have been reviewing their operations to minimize THM formation without compromising public health protection. This has involved adjustment to the type and amount of chlorine used as well as where it is applied. In addition, the treatment process has been expanded to remove the naturally occurring organic matter that reacts with chlorine to produce THMs.

Other means of disinfection besides chlorine are available. However, these methods may also produce harmful by-products. In addition, alternative disinfectants cannot provide the residual protection (that is, continue to disinfect in the city-wide distribution system) of chlorine-based disinfectants, so they must be used in combination with chlorine.

Drinking water treatment operations must often meet competing objectives—adequate microbial protection, reduced levels of disinfection by-products, and corrosion control—to comply with EPA regulations. The key to treatment is to provide a balance between the health benefits of disinfected drinking water and the creation of by-products from the disinfectants.

It is not an easy task and is one that requires close and continuous attention. It is an ongoing process, and EPA is continuing to revise its regulations to provide the balance to prevent by-products with long-term health effects while keeping the microbiological quality of drinking water as the top priority.

Vocabulary

by-products	副产物	residual	残余的,残留的
awareness of	意识到	long-term	长期的
Environmental Protection Agency	环保局	ongoing	正在进行的

Unit 15 Chemical Precipitation Softening

The process of removing hardness from water is called softening. Hardness is mainly caused by the presence of calcium and magnesium salts. These substances are dissolved from deposits through which the water percolates. The length of time that water is in contact with the hardness-producing material is one factor that determines how much hardness there is in the raw water. To soften water by the lime or soda-ash method, its degree of alkalinity has to be considered.

The alkalinity of a water sample is a measure of the water's capacity to neutralize acids. In natural and treated waters, alkalinity is the result of the presence of bicarbonates, carbonates, and hydroxides of calcium, magnesium, and sodium. Many of the chemicals used in water treatment, such as alum, chlorine, or lime, cause changes in alkalinity. Determining of alkalinity is useful when calculating chemical dosages for the coagulation and water-softening processes. The alkalinity must also be known when calculating corrosivity of the water and to estimate its carbonate hardness. Alkalinity is usually expressed in terms of calcium carbonate:

Alkalinity = bicarbonate ion concentration [HCO_3] + carbonate ion concentration [CO_3] + hydroxyl ion concentration [OH], expressed as calcium carbonate $CaCO_3$.

The two basic methods of softening public water supplies are chemical precipitation and ion exchange. Other methods can also be used to soften water, such as electrodialysis, distillation, freezing, and reverse osmosis. These processes are complex and expensive and usually used only in unusual circumstances.

Chemical precipitation is one of the more common methods used to soften water. The chemicals normally used are lime (calcium hydroxide, $Ca(OH)_2$) and soda ash (sodium carbonate, Na_2CO_3). Lime is used to remove the chemicals that cause the carbonate and magnesium non-carbonate hardness. Soda ash is used to remove the chemicals that cause the non-carbonate hardness.

When lime and soda ash are added, the hardness-causing minerals form nearly insoluble precipitates. When calcium hardness is removed in a chemical softener, it is precipitated as calcium carbonate ($CaCO_3$). When magnesium harness is removed in a chemical softener, it is precipitated as magnesium hydroxide ($Mg(OH)_2$). These precipitates are removed by the conventional processes of coagulation/flocculation, sedimentation, and filtration. Because the precipitates are very slightly soluble, some hardness remains in the water—usually about 50 to 85 mg/L (as $CaCO_3$). This hardness level is desirable to prevent corrosion problems associated with water being too soft and having little or no hardness ions.

Conventional Lime-Soda Ash Treatment

When the content of the water has little magnesium hardness only the calcium ions need to be removed. This is termed conventional lime-soda treatment. Only enough lime and soda ash are added to the water to raise its pH to between 10.3 and 10.6. The calcium hardness will then be removed from the water, but little if any magnesium hardness will be removed.

Excess Lime Treatment

When the magnesium hardness of water is more than about 40 mg/L (expressed as $CaCO_3$), magnesium hydroxide scale will deposit in household hot-water heaters operated at normal temperatures of 140 to 150 EF. To reduce this magnesium hardness, more lime than is used in the conventional process must be added to the water. The extra lime will raise the pH above 10.6 so that the magnesium hydroxide will precipitate out of the water. This process is known as excess lime treatment.

Split Treatment

When the water contains high amounts of magnesium, a process called split treatment may be used. Approximately 80 percent of the water is treated with an excess of lime to remove the magnesium at a pH above 11, after which it is blended with 20 percent of the source water. Split treatment can reduce the amount of carbon dioxide required to recarbonate the water as well as offer a savings in lime feed. Since the fraction of the water that is treated contains an excess lime dose, magnesium is almost completely removed from this portion. When the water is mixed with the water that does not undergo softening, the carbon dioxide and bicarbonate in that water tend to recarbonate the final blend.

The split treatment process reduces the amount of chemical needed to remove the hardness from the water by about 20 to 25 percent, a significant saving.

Design Considerations

In lime soda-ash softening plants, the softening process may be carried out by a sequence of rapid mix, flocculation, and sedimentation or in a solids-contactor softener. In the solids contactor the rapid mix, flocculation, and sedimentation occur in a single unit. The process begins with the mixing of the chemicals into the water, followed by violent agitation, termed rapid mixing. This allows the chemical to act on the hardness in the water to precipitate the calcium or magnesium hardness in the water.

The purpose of the flocculation step is to allow the flocs to contact other flocs and grow large enough to settle in the sedimentation stage. During this step the water is mixed gently with a small amount of energy expended in the mixing. Most flocculators are now compartmentalized, allowing for a tapered mix, so that less energy needs to be applied as the flocs grow in size.

The retention time in the flocculator is important to allow the particles to come in contact with each other. The minimum time recommended is 30 minutes for conventional water softening. Sludge returned to the head of the flocculator reduces the amount of chemical needed and provides seed flocs for the precipitation. The estimated return sludge is 10 to 25 percent of the source water.

Sedimentation is the next step following flocculation. Settling rates for the tanks are a function of the particle size and the density. The detention times in the settling basins range from 1.5 hours to 3.0 hours. The basins can be of many shapes. They can be rectangular, square, or circular; some designs incorporate inclined tube settlers.

Another type of sedimentation basin is the solids-contact unit, in which the water is mixed with chemicals and flocculated in the center of the basin, then forced down and trapped for removal in a sludge blanket in the bottom of the tank.

Sludge Removal

The residue created from a water-softening process done with lime and soda ash is normally very high in calcium carbonate or a mixture of calcium carbonate and magnesium hydroxide. Calcium carbonate sludges are normally dense, stable and inert and dewater readily. The solids content in the sludge will range from 5 percent to 30 percent total solids. The pH of the sludge will be greater than 10.5.

Water plant sludges may be treated by many processes, such as lagooning, vacuum filtration, centrifugation, pressure filtration, recalcination, or land application. The most common method is the storage of sludge in lagoons and the application to farmland or disposal in landfills.

Vocabulary

softening	软化	electrodialysis	电渗析
hardness	硬度	centrifugation	离心
magnesium	镁	pressure filtration	压滤
alkalinity	碱度	farmland	农田
bicarbonate	重碳酸盐	landfill	(垃圾)填埋

Reading Material A

Recarbonation

Depending on the amount of lime or chemical needed to reduce the amount of calcium or magnesium in the water, the treated water will generally have a pH greater than 10. It is necessary to lower the pH to stabilize the water and prevent the deposition of hard carbonate scale on filter sand and distribution piping. Recarbonation is the most common process used to reduce the pH. This procedure involves the addition of carbon dioxide to the water after the softening. Generally, enough carbon dioxide is added to reduce the pH of the water to less than 8.7. The actual amount of carbon dioxide to add must be determined by using a saturation index of some kind. The Langelier Index (LI) is by far the most common stabilization index used, but some plants instead use the Rizner Index, the reciprocal of the Langelier Index. The Langelier Index is expressed as the pH of stabilization (pHs) minus the actual pH measured (pHs-pH). When the Langelier Index is positive, the water will tend to coat the pipes. When it is negative, the water tends to be corrosive.

Two types of recarbonation processes are used with the four types of softening processes. Because of the high capital cost for building the two-stage treatment train, single-stage recarbonation is the one most commonly practiced.

Because of the high capital cost for building the two-stage treatment train, single-stage recarbonation is the one most commonly practiced. There are some benefits to using the two-stage method, such as reduced operating cost since less carbon dioxide is needed. Also, better finished water quality is usually obtained through the two stage process.

Single Stage Recarbonation

For treatment of low-magnesium water, where excess-lime addition is not required, single-stage recarbonation is used. The water is mixed with lime or soda ash in the rapid-mix basin, resulting in a pH of 10.2 to 10.5. If non-carbonate hardness removal is required, soda ash will also be added at this step. After rapid mixing, the resulting slurry is mixed gently for a period of 30 to 50 minutes to allow the solids to flocculate. After flocculation, the water is allowed to flow into a sedimentation basin where the solids will be removed by sedimentation. Following sedimentation the clear water flows to the recarbonation basin where carbon dioxide is added to reduce the pH to between 8.3 and 8.6. Any particles remaining in suspension after recarbonation are removed by filtration.

Two-stage Recarbonation

Two-stage softening is sometimes used to recarbonate high magnesium water that has to be treated with excess lime. In this process, excess lime is added in the first stage to raise the pH to 11.0 or higher for magnesium removal. Following first stage treatment, carbon dioxide is added to reduce the pH to between 10.0 and 10.5, the best value for removal of calcium carbonate. If non-carbonate hardness removal is required, soda ash will be added at this point. After second stage treatment, the water flows to a secondary recarbonation tank, where the pH is reduced to between 8.3 and 8.6.

Vocabulary

recarbonation	再碳化	saturation index	饱和指数
carbon dioxide	二氧化碳	slurry	浆,泥浆

Reading Material B

Ion-exchange Softening

The two basic methods of softening public water supplies are chemical precipitation and ion exchange. Other methods can also be used to soften water, such as electrodialysis, distillation, freezing, and reverse osmosis. These processes are complex and expensive and usually used only in unusual circumstances.

The ion-exchange method of water softening has been used extensively in smaller water systems and individual homes. It is based on the ability of the ion-exchange resin zeolite to exchange one ion from the water being treated with another ion in the resin. Zeolite resin exchanges sodium ions for ions causing hardness in the water. Such ions are the ions of calcium and magnesium present in the water that flows through the zeolite resin bed. The water treated ends up with more sodium ions than before, which could be a potential problem for persons who must watch their salt intake. Sodium is one component of salt, chlorine being the other component.

The ion-exchange softening process does not alter the water's pH or alkalinity. However, the stability of the water is altered due to the removal of the calcium and magnesium and an increase in dissolved solids. For each mg/L of calcium removed and replaced by sodium, the total dissolved solids increases by 0.15 mg/L. For each mg/L of magnesium removed and replaced by sodium, the total dissolved solids will increase by 0.88 mg/L.

The measurements used to express water hardness in the ion-exchange process differ from the units that are normally used in the lime-soda softening process. Hardness in the ion-exchange process is expressed as grains per gallon rather than mg/L of calcium carbonate. The following conversions show the relationship between mg/L and grains per gallon:

$$1 \text{ grain} = 17.12 \text{ mg/L}$$

$$1 \text{ grain} = 0.143 \text{ pound per 1000 gallon}$$

$$7\,000 \text{ grains} = 1 \text{ pound per gallon}$$

An example of this relationship: if a water contains 10 grains per gallon, what would the hardness be expressed as in mg/L?

$$10 \text{ grains per gallon} \times 17.12 \text{ mg/L} = 171.2 \text{ mg/L of hardness}$$

Advantages of the Ion-exchange Process

Compared with lime-soda ash softening, ion-exchange softening has certain advantages, the main ones being its compactness and its low cost. The chemicals used are safer for the operator to handle, and the operation of the zeolite-softening process is much easier. It can be almost totally automated. As the resins have the ability to remove all of the hardness from the water, the treated water must be blended with untreated water to obtain the hardness level the operator would like to maintain.

Many utilities have found the ion-exchange process to be the most cost effective way of producing quality water for their customers. If the zeolite units are used to soften surface water, it has to be preceded by the normal surface-water treatment units of rapid mix, flocculation, sedimentation, and filtration. This is needed to remove any material that may otherwise foul the resin. Therefore, the lime-soda process is most likely the more cost-effective method for surface-water softening than the ion-exchange one.

Ion Exchange Material

There are a number of materials, including some types of soils, that can act as cation exchangers for softening. A natural green sand called glauconite has very good exchange capabilities, and it was once widely used. Synthetic zeolites, known as polystyrene resins, are the ones most commonly used now. Their cost is reasonable, and it is easy to control the quality of the resin. They also have much higher ion-exchange capacities than the natural material.

The ability of the resin to remove hardness from the water is related to the volume of resin in the tank. The zeolite resin should remove about 50,000 grains of hardness per cubic foot of resin. The resin holds the hardness ions until the tank is regenerated, when the hardness ions are exchanged for sodium ions in the salt brine used as the regenerating solution.

Ion Exchange Units

The units containing the resin resemble pressure filters. The inside is generally treated to protect the tank against corrosion from the salt. The units are normally of the downflow type, so the media also acts as a filter. The size and volume of the units are dictated by the hardness of the water and the volume of treated water needed to be produced between each regeneration cycle. The resin is supported by an underdrain system that removes the treated water and distributes the brine evenly during the regeneration process. The minimum depth of the resin should be no less than 0.61 m above the underdrain.

Salt Storage

The salt is stored as a brine, ready to be used for the regeneration of the resin. The amount of salt needed ranges from 0.25 to 0.45 pounds for every 1,000 grains of hardness removed. The tank should be coated with a salt-resistant material to prevent corrosion of the tank walls.

Brine-feeding Equipment

Concentrated brine contains approximately 25 percent salt. The brine should be diluted to about 10 percent before it is added to the tank during regeneration. It is generally injected with the use of a venturi or by the use of a metering pump.

The solubility of salt decreases with a rise in temperature, which forces the salt out of solution. The water that remains after the salt has separated out of the solution is subject to freezing. Therefore, the brine piping should be protected from cold temperatures.

Devices for Blending

A properly operated ion-exchange unit produces a water with zero hardness but with high corrosivity. Since a total hardness of 85 to 100 mg/L is the most desirable, the effluent from the ion-exchange unit is generally blended with source water to raise the hardness in the finished water. Blending is normally accomplished by metering both the effluent from the softener and the raw water added. Meters are installed in both lines so that the operator can adjust and monitor the blend.

Operation of Units

The basic steps in the operation of an ion-exchange softening unit are the softening cycle, backwash, regeneration, slow rinse, and fast rinse.

Softening Cycle The softening cycle involves the feeding of water into the unit until hardness appears in the effluent from the unit. The cycle ends when 1 to 5 mg/L of hardness is detected in the effluent. Almost all softening units have an alarm on the water meter to indicate when a certain amount of water has passed through the exchange unit. Loading rates for synthetic resins are in the area of 24.4 to 36.6 m/h of media surface area.

Backwash Cycle Once the hardness breaks through, the softener unit needs to be regenerated. In down-flow units the resin must first be backwashed to loosen the resin, since it becomes compacted by the weight of the water, and to remove any material that has been filtered out of the water by the resin. The backwash rate is normally 14.64 to 19.52 m/h of zeolite bed area. The operator needs to apply enough backwash water to expand the resin bed by about 50 percent. The backwash water is usually discharged to a box containing orifice plates that establish and measure the flow rate. Distributors at the top of the unit provide for uniform water distribution and uniform wash-water collection. Underdrains help the uniform distribution of the backwash water on the bottom of the resin.

Regeneration To regenerate an ion-exchange unit, concentrated brine is pumped to the unit from the storage basin. The brine is diluted through the injector to a solution containing about 10 percent salt before it is passed through the resin. The time required for regeneration is about 20 to 35 minutes. The flow rate of brine through the resin is measured in gallons per minute per cubic foot of media.

The brine needs to be in contact with the resin long enough to allow for complete exchange of the hardness ions in the resin with sodium ions in the brine. It is better to allow too much time than to not allow enough. If the resin is not totally recharged, the next softening run will be short.

Vocabulary

ion exchange	离子交换	compactness	紧密，简洁
resin	树脂	glauconite	海绿石，海绿沙
zeolite	沸石	regenerate	再生

PART TWO

Unit 16 Surface Water Pollution

Every human activities affecting watershed components can have a strong impact on the surface water pollution. One way people influence surface water composition is by adding potential pollution sources to the watershed. How the land in a watershed is used by people, whether it is farms, houses or shopping centers, has a direct impact on the surface water quality collected from the watershed. When it rains, stormwater carries with it the effects of human activities as it drains off the land into the local waterway. As rain washes over a parking lot, it might pick up litter, road salt and motor oil and carry these pollutants to a local stream. On a farm, rain might wash fertilizers and soil into a pond. Snow melt might wash fertilizers and pesticides from a suburban lawn.

Sources of Surface Pollution

As for ground water, surface water pollution can derive from point sources and distributed non-point sources. However the degree of pollution or the pollution risk of surface water is far higher. Runoff over a watershed or punctual discharges of industrial and urban wastewater determine an immediate and direct contact between pollutants and superficial water bodies which consequently attain high concentrations of most all type of pollutants. In the case of groundwater only some contaminant (the more soluble ones) are of serious concern, as soil and rock layers above aquifer act as an active filter for most of the compounds dissolved or suspended.

Therefore while for ground water the pollution problem, even if of great hazard, is limited to dissolved mineral (such as nitrate and chloride), to the most soluble heavy metals and to some organic compounds (such as pesticides), there are no limits to the type and to the concentration of substance which can enter surface waters.

Surface Water Pollutants

The main group of pollutants which can affect surface water bodies are: 1) organic material, 2) phosphorus, 3) heavy metals, 4) detergents and surfactants, and 5) trace synthetic organic compounds.

Organic material Organic material can be found in surface waters either as dissolved compounds or in a suspended form. It can be measure with TOC analyzers or indirectly by the determination of BOD, BOD_5 and COD.

Organic matter in used from water organisms either as energy source or for producing growth substances. Biodegradation of organic compounds requires oxygen as the energy production is an oxidation process requiring oxygen as final electron acceptor. Therefore level of organic matter in the water bodies can cause a rapid oxygen depletion, especially in condition of low oxygen turnover (high temperature, still water). The decrease of dissolved oxygen impairs the growth of aerobic microorganism which decompose organic matter producing CO_2, H_2O, nitrates, sulphates and phosphates as final products and favours anaerobic conditions which are detrimental for the aquatic life. In fact, anaerobic microflora is responsible for putrefactive processes which are characterized by the production of different types of toxic and noxious compounds (ammonia, hydrogen sulfide and phosphine) as final products of the organic matter degradation.

Phosphorus Phosphorus can enter the surface water in form of phosphate and of organic compounds. It can derive either from distributed sources or from point source. The most important diffuse source is rain water runoff from natural and agricultural soil. Point source of phosphorus pollution are black water (containing organic phosphorus from human metabolism) and gray water from urban and rural settlements, industrial waste water and wastewater from animal husbandry.

Agricultural role in phosphorus pollution seem to be of scarce relevance. Fertilizer phosphate is tightly adsorbed by soil colloids and only phosphate in soil particle can reach surface water as consequence of erosive processes. But also in this case the polluting impact is thought to be minimal as it has been shown that algae cannot utilize absorbed P but only soluble inorganic phosphate. Only when dissolved phosphate concentration is very low a passage of a adsorbed phosphate into the solution occurs. Therefore the presence of soluble phosphate and organic P compounds in surface water is due to the urban wastewater discharge, to some industrial wastewater and to the discharge of water from intensive animal farms.

The main important effect of P on surface water is eutrophication. As P is the limiting factor for algal growth when high concentration of this element are attained as impressive algal bloom can occur. The considerable amount of organic material deriving from the alga photosynthetic process is decomposed from water aerobic microorganism thus depleting oxygen and triggering anaerobic putrefactive conditions. Oxygen deficiency and toxic substance from anaerobic metabolism cause fish death and impairment of aquatic life. Moreover in the case of lake and reservoirs with a long time of water turnover phosphorus will accumulate in the aquatic ecosystem determining periodic cycles of algal proliferation with inorganic P being organized in the algae cells followed by microbial decomposition of algal residues with the organic P being remineralized. Only the removal of organic substance from the lake either as sludge accumulated on the bottom or as living organism (e.g. fish) can reduce the water body eutrophication.

Basic on increasing nutrient load, water body can be classified as oligotrophic, mesotrophic and eutrophic. According to the OCSE guidelines oligotrophic water have a total P concentration lower than 10 micrograms per liter, while in eutrophic water total P concentration is higher than 50 micrograms per liter. However other parameters as for example water transparency, chlorophyll concentration and BOD should be considered for estimating the eutrophication degree of a water body.

Heavy metals Heavy metals (such as mercury, lead, chromium and cadmium) may originate in industrial discharges, runoff from city streets, mining activities, leachate from landfills, and a variety of other sources. Also agriculture can contribute to heavy metal pollution as these elements are contained in some pesticides and as trace in fertilizer. Most of these metals which are generally persistent in the environment are hazardous for any aquatic ecosystem as well as for human health. The polluting potential of heavy metals depends not only on their concentration in water but also on the form in which they are present. It is well known that with the exception of mercury the toxicity of heavy metals is due to the ionic form while the coordinated and the precipitated forms are less hazardous. Therefore all the conditions favouring the formation of heavy metal ions (e.g. for many metals low pH values) increase the risk of water contamination. On the other hand it must be considered that precipitation of heavy metals or their absorption on suspended and sedimented particles poses a long term risk as heavy metal ions can be liberated if conditions favorable to solubilization occur.

The hazardousness of heavy metals is increased from their accumulation in some organism of the food chain, like for example mollusks and algae.

Therefore acceptability limits of heavy metals concentrations have been proposed in order to protect some aquatic species and to avoid bioaccumulation phenomena.

Detergents and Surfactants Surfactants and detergents are common contaminants of surface water due to their large consume and to their use for every type of washing and cleaning operation. These synthetic compounds are characterized by a hydrophilic head and a lipophilic tail consisting of a C-based linear or branched chain. The group constituting the hydrophilic head, which is responsible for the water solubility of detergents can be anionic, cationic or polar. The anionic detergents like fatty acids salts (soap) and the alchilbenzensulfonate salts are the most common products used in the every day washing.

Detergents and surfactants have a high biodegradability and have a moderate toxicity for fish. However they can affect the oxygen turnover of surface water and reduce the sedimentation process thus delaying water clarification. The threshold values which can impair aquatic life are 3-12 mg/L for anionic detergents, 3-38 mg/L for non-ionic detergents and 20 for cationic detergents.

Trace Synthetic Organic Compounds Many synthetic organic compound can be found as trace contaminants in surface water. Even if they are present in very low concentration they can represents a long term hazard due to accumulation process either in food chain or in water sediment. Source of these organic compounds can be the agricultural activities (e.g. pesticide use), accidents (like the oil pipeline breakage) and the industrial discharges. Wastewater deriving from chemical industries are the main source of surface water contamination with synthetic organic compounds, the type of polluting varying according the type of processed material. Factor affecting the toxicity of synthetic organic compound are molecular structure, water solubility, biodegradability, photodegradability and volatility.

Halogen containing compound are retained the most dangerous ones due to their persistence (resistance to biodegradation) and to their accumulation. Hazardousness of halogenated compounds seems to increase with the increase of the number of halogen atoms in the molecule.

Soluble chemicals are generally less dangerous than insoluble or fatty soluble substances as they scarcely accumulate in organisms, being removed by the regular turnover mechanisms. Also their accumulation in water sediments and on water particles is reduced.

Biodegradable organic chemicals like detergents and surfactant show a low toxicity risk as they are decomposed by water microorganism and are therefore less susceptible to the accumulation process.

Volatile substance have also a reduced contamination potential for surface water as they tend to leave the liquid phase. Their presence in water is also influenced from the water movement which favour the aeration process.

Also the susceptibility of organic compounds to light degradation generally limits their polluting effect, unless toxic substance are produced by the decomposition process.

Vocabulary

fertilizer	肥料	husbandry	饲养
pesticide	杀虫剂	eutrophication	富营养化
suburban	郊外的,偏远的	trigger	引发反应,引起
soluble	可溶的,可溶解的	oligotrophic	贫营养的
detergent	清洁剂,去垢剂	mesotrophic	中度营养的
surfactant	表面活性剂	mollusks	软体动物
synthetic	合成的,人造的	hydrophilic	亲水的
biodegradation	生物降解	lipophilic	亲脂的
aerobic	好氧的	alchilbenzensulfonate	烷基苯磺酸
anaerobic	厌氧的	photodegradability	光降解性
putrefactive	腐败的,腐化的		

Reading Material A

Ground Water Contamination

Nearly all public water supplies in Montana provide safe drinking water. Incidents of ground water contamination, however, have been reported in several areas of the state.

Ground water can become contaminated from natural sources or numerous types of human activities. Residential, municipal, commercial, industrial, and agricultural activities can all affect ground water quality. Contaminants may reach ground water from activities on the land surface, such as industrial waste storage or spills; from sources below the land surface but above the water table, such as septic systems; from structures beneath the water table, such as wells; or from contaminated recharge water.

The following are examples of how wells themselves can cause contamination of a source of water supply:

1. Improperly abandoned wells act as a conduit through which contaminants can reach an aquifer if the well casing has been removed, as is often done, or if the casing is corroded. In addition, some people use abandoned wells to dispose of wastes such as used motor oil. These wells may extend into an aquifer serving drinking water supply wells.

2. Active drinking water supply wells which are poorly constructed can allow conditions which may result in ground water contamination. Construction problems such as faulty casings, inadequate well caps, lack of grouting, or lack of adequate drainage away from the well may allow outside water and any accompanying contaminants to flow into the well. Sources of such contamination can be surface runoff or wastes from farm animals or septic systems. Contaminated fill packed around a well can also degrade well water quality. Well construction problems are more likely to occur in wells that were in place before the establishment of well construction standards.

3. Poorly constructed irrigation wells also can allow contaminants to enter ground water. Pesticides and fertilizers may be inappropriately applied in the immediate vicinity of wells on agricultural land.

Ground water contamination may be biological or chemical or radiological in nature. In any case, contamination often remains undetected until monitoring reveals the presence of the contaminant. In only a very few instances was contamination in Montana public water supply systems detectable by taste or odor of the water. In these cases the amount of the chemicals present far exceeded standards established to protect public health.

Contamination of ground water can result in poor drinking water quality, loss of a water supply, high cleanup costs, high costs for alternative water supplies, and potential liability or health problems.

Biological Contaminants

Possible biological contaminants include bacteria, viruses, and parasitic protozoans. The primary concern over biological contamination is that the organisms may be pathogenic, which means they are capable of causing disease. Often the waterborne diseases result in gastrointestinal illness. In some cases untreated symptoms may result in death of the infected person.

In Montana, outbreaks of typhoid fever and cholera in the early 1900's led to the first state drinking water regulations. In the late 1970's and early 1980's, outbreaks of giardiasis in public water supply systems using surface water in Missoula, Red Lodge and White Sulfur Springs led to more stringent treatment requirements for surface water.

It is not possible to analyze water samples for every potential biological contaminant which may be present. As a result, detecting biological contamination of a public water supply system largely depends on ensuring the system is properly constructed and operated, and on monitoring the water for indicator bacteria in representative water samples. Test methods for parasites *Giardia* and *Cryptosporidium* and for viruses are improving, but are not sufficiently sensitive or affordable to warrant their routine use as a monitoring tool. However, in certain circumstances monitoring for them may be required to investigate a suspect source or public water supply system.

Bacteria are single-celled organisms occurring in the environment, on our skin and in our bodies. Some are essential to our survival and some may make us sick. Waterborne diseases caused by bacteria include, among others, cholera, Legionnaire's Disease and gastrointestinal illness caused by a particular strain of E. Coli—one of the members of the fecal coliform group.

Viral disease agents such as the Norwalk virus and hepatitis may also be waterborne. To cause infections in humans, viruses must come from sources of pollution such as human fecal contamination or septic tank system effluent. Cross-connections to nonpotable water sources have been tied to some viral disease outbreaks.

Two parasitic protozoans of concern in water are *Giardia lamblia* and *Cryptosporidium*. These multi-celled organisms come from the intestinal tracts of humans and a variety of animals. They are very resistant to disinfection and generally must be removed from water through filtration treatment. They can contaminate a ground water source if it is ground water under the direct influence of surface water, if septic system discharges are recharging ground water within the zone of contribution, or if the source is poorly constructed.

Chemical Contaminants

Potential chemical contaminants of public water supply systems include naturally occurring inorganic chemicals as well as organic compounds such as solvents and pesticides.

In Montana, gasoline derivatives leaking from underground storage tanks, improper disposal of solvents used in the dry cleaning process, and weed-killers have been responsible for chemical contamination of ground water sources. Chemical contaminants may cause a variety of health effects including gastrointestinal illness, cancer risk, nervous system effects, or effects on internal organs. Monitoring for chemical contaminants differs from biological contaminants in that specific chemicals can be individually identified in water samples.

Some chemical substances found naturally in rocks or soils, such as arsenic, iron, manganese, chloride, fluoride, or sulfate can become dissolved in ground water. Naturally occurring nitrate occurs in some areas of Montana as do relatively high levels of arsenic and fluoride. A wide variety of potential sources of chemical contamination of ground water exist. Some septic system cleaners and additives, improper disposal of household cleaners, improper chemical storage, sloppy materials handling, and poor quality containers can be major threats to ground water. Chemical contamination incidents in Montana have occurred through leaking underground fuel tanks, improper disposal of cleaning solvents, and inadvertent contamination of shallow aquifers by flushing solvents down floor drains connected to septic tanks or dry wells. Sewer pipes carrying wastes have also been found to sometimes leak contaminated fluids into the surrounding soil and ground water.

Radiological Contaminants

Radioactivity in the form of radium, uranium, and radon gas naturally occurs in ground water in some parts of the United States. Montana has a significant number of aquifers with radon at relatively high levels. Public water supply systems with radioactive contaminants have options for blending to achieve water below the applicable MCL, abandoning the source or applying treatment to reduce the contaminant to an acceptable level.

Vocabulary

contaminant	污染物	giardiasis	贾第鞭毛虫病
radiological	放射性的	giardia	贾第虫
protozoan	原生动物	cryptosporidium	隐孢子虫

Reading Material B

Cause and Sources of Surface Water Pollution

The major causes of pollution in surface water in Texas are high fecal coliform levels, low dissolved oxygen, metals or other toxics such as organics in water and elevated concentrations of dissolved solids. Overall, in Texas, 59 water body segments have been impacted by high fecal coliform levels in water or shellfish, while low dissolved oxygen has impacted 33 segments. Other causes of impairment include high metal content (33 segments), high organics in fish and shellfish (36), and elevated concentrations of dissolved solids (17).

However, each type of water body is impacted by different types of impairment. Major contributors to nonsupport of uses in bays included elevated fecal coliform bacteria and low dissolved oxygen contents. In the streams and rivers, the most frequently violated water quality standards were those for pathogens (high levels of fecal coliform bacteria), low dissolved oxygen, and, in some areas, toxics such as metals and pesticides. Excess plant nutrients (nitrogen and phosphorus) also were identified as a problem in some waters of the state. In reservoirs, impairment of use was related to elevated levels of metals and high levels of pathogens and pesticides (see Fig. 16.1).

According to the 1996 Water Quality Inventory and data drawn from the state's limited monitoring of toxics, 839.2 km of streams and rivers, 22,240 acres of reservoir, and 1.1 km of bays and estuaries have toxicity levels so high they do not meet their use for aquatic life. In addition, the TNRCC has identified 21 segments of concern for toxic substances in ambient water, as well as 18 segments where at least two tests evidenced in-stream toxic effects from water or sediment. Ambient water toxicity is due to metals such as cadmium, zinc, lead, silver, and aluminum, while diazinon, an organic pesticide, has exceeded both acute and chronic screening levels in two water segments. Finally, eight segments have been identified as being of concern for toxic substances in fish tissue.

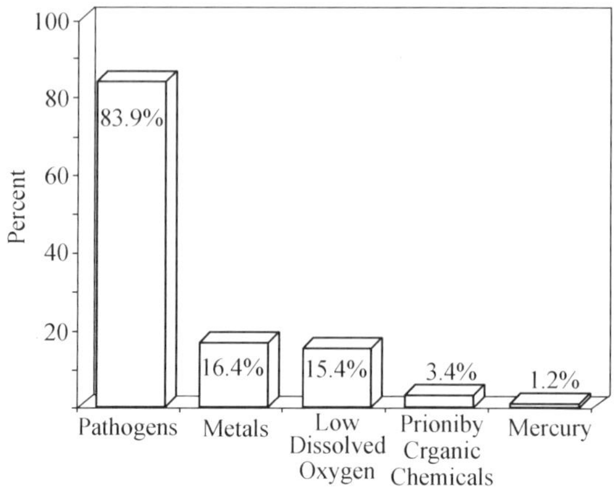

Fig. 16.1 Causes present in rivers and streams (% of miles), 1998

Others include: Pesticides—1.0%; Salinity/TDS/Chlorides—0.6%; Nutrients—0.6%

The sources of water pollution typically fall into one of two categories: point-source pollution and non-point-source pollution. The term point-source pollution refers to pollutants discharged from one discrete location or point, such as an industry or municipal wastewater treatment plant. Pollutants discharged in this way might include, for example, fecal coliform bacteria and nutrients from sewage, and toxics such as heavy metals, or synthetic organic contaminants.

The term non-point-source pollution refers to pollutants that cannot be identified as coming from one discrete location or point. Examples are oil and grease that enter the water with runoff from urban streets, nitrogen from fertilizers and pesticides, and animal wastes that wash into surface waters from agricultural lands. Natural and unknown causes of pollutants also can impact water quality and may be related to human activities. For example, highway or housing construction may help precipitate the runoff of natural pollution sources, such as sediment. Sources present in surface water see Tab. 16.1.

Tab. 16.1 Sources present in surface waters, 1998

Rivers and Streams		Lakes and Reservoirs	
Source	% of Miles	Source	% of Acres
Municipal Point Souces	48.7	Atmospheric Deposition	8.3
Agriculture	25.2	Municipal Point Sources	4.1
Urban Runoff/Storm Sewers	18.1	Agriculture	2.9
Range Grazing	8.7	Urban Runoff/Storm Sewers	2.7
Industrial Point Sources	5.2	Natural Sources	2.2
Natural Sources	4.4	Industrial Point Sources	1.5
Animal Feeding Operations	3.6	Range Grazing	1.5
Land Disposal	1.3	Land Disposal	1.4
Pasture Grazing	1.3	Hydromodification	0.02
Flow Modifications	0.8		
Irrigated Crop Production	0.5		
Septic Tanks	0.4		

Vocabulary

segment	段,节,分割	zinc	锌
reservoir	水库,蓄水池	silver	银
coliform	大肠菌	diazinon	二嗪农(用作杀虫剂)
ambient	周围的	chronic	慢性的
estuary	河口		

Unit 17 Constituents in Wastewater

An understanding of the nature of wastewater is essential in the design and operation of collection, treatment, and reuse facilities, and in the engineering management of environmental quality. To promote this understanding, the information in this lesson is presented in nine sections dealing with 1) physical characteristics, 2) inorganic constituents, 3) organic constituents, 4) and biological characteristics.

Physical Characteristics

The most important physical characteristic of wastewater is its total solids content, which is composed of floating matter, settleable matter, colloidal matter, and matter in solution. Other important physical characteristics include particle size distribution, turbidity, color, transmittance, temperature, conductivity, density and odor. Wastewater contains a variety of solid materials varying from rags to colloidal material. In the characterization of wastewater, coarse materials are usually removed before the sample is analyzed for solids. Typically, about 60 percent of suspended solids in a municipal wastewater are settleable. Total solids (TS) is obtained by evaporating a sample of wastewater to dryness and measuring the mass of the residue. Turbidity, a measure of the light-transmitting properties of water, is another test used to indicate the quality of waste discharges and natural waters with respect to colloidal and residual suspended matter. In general, there is no relationship between turbidity and the concentration of total suspended solids in untreated wastewater. There is, however, a reasonable relationship between turbidity and total suspended solids for the settled and filtered secondary effluent from the activated sludge process.

Chemical Characteristics

The chemical constituents of wastewater are typically classified as inorganic and organic.

Inorganic constituents Inorganic chemical constituents of concern include nutrients, nonmetallic constituents, metals, and gases. The sources of inorganic nonmetallic and metallic constituents in wastewater derive from the background levels in the water supply and from the additions resulting from domestic use, from the addition of highly mineralized water from private wells and groundwater, and from industrial use. Nonmetallic constituents include pH, nitrogen, phosphorus, alkalinity, chlorides, and sulfur.

Trace quantities of many metals, such as cadmium, chromium, copper, iron, lead, manganese, mercury, and zinc are important constituents of most waters. Many of these metals are also classified as priority pollutants. However, most of these metals are necessary for growth of biological life, and absence of sufficient quantities of them could limit growth of algae, for example. The presence of any of these metals in excessive quantities will interfere with many beneficial uses of the water because of their toxicity; therefore, it is frequently desirable to measure and control the concentration of these substances.

Organic constituents Organic compounds are normally composes of a combination of carbon, hydrogen, and oxygen, together with nitrogen in some cases. The organic matter in wastewater typically consists of proteins, carbohydrates, and oils and fats. Over the years, a number of different analyses have been developed to determine the organic content of wastewaters. In general, the analyses may be classified into aggregate and individual. Aggregate organic constituents are comprised of a number of individual compounds that cannot be distinguished separately. Both aggregate and individual organic constituents are of great significance in the treatment, disposal, and reuse of wastewater. In general, the analyses used to measure aggregate organic material may be divided into those used to measure gross concentrations of organic matter. Laboratory methods commonly used today to measure gross amounts of organic matter in wastewater include: 1) biochemical oxygen demand (BOD), 2) chemical oxygen demand (COD), 3) total oxygen demand (TOD), 4) total organic carbon.

Individual organic compounds are determined to assess the presence of priority pollutants and a number of new emerging compounds of concern. As the techniques used to identify specific compounds continue to improve, a number of other organic compounds have been detected in public water supplies and in treated wastewater effluents.

Biological Characteristics

Organisms found in surface water and wastewater includes bacteria, fungi, algae, protozoa, plants, animals, and viruses. Bacteria, fungi, algae, protozoa, and viruses can only be observed microscopically. The biological characteristics of wastewater are of fundamental importance in the control of disease caused by pathogenic organisms of human origin, and because of the extensive and fundamental role played by bacteria and other microorganisms in the decomposition and stabilization of organic matter, both in nature and in wastewater treatment plants.

It should be noted that many of the physical properties and chemical and biological characteristics are interrelated. For example, temperature, a physical property, affects both the amounts of gases dissolved in the wastewater and the biological activity in the wastewater.

The important constituents of concern in wastewater treatment are suspended solids, biodegradable organics, pathogens, nutrients, priority pollutants, heavy metals and dissolved inorganic. Secondary treatment standards for wastewater are concerned with the removal of biodegradable organics, total suspended solids, and pathogens. Many of the more stringent standards that have been developed recently deal with the removal of nutrients, heavy metals, and priority pollutants. When wastewater is to be reused, standards normally include additional requirements for the removal of refractory organics, heavy metals, and in some cases, dissolved inorganic solids.

Vocabulary

particle size distribution	粒径分布	iron	铁
colloidal matter	胶体物质	lead	铅
transmittance	透光度	manganese	锰
conductivity	传导性	mercury	汞
priority pollutant	优先控制污染物	zinc	锌
interfere with	干涉,影响	characteristic	特性特征
cadmium	镉	protein	蛋白质
chromium	铬	fungi	真菌
copper	铜	pathogenic	致病的

Reading Material A

Impact of Regulations on Wastewater Engineering

From about 1900 to the early 1970s, treatment objectives were concerned primarily with 1) the removal of colloidal, and floatable material, 2) the treatment of biodegradable organics, and 3) the elimination of pathogenic organisms. Implementation in the United States of the Federal Water Pollution Control Act Amendments of 1972 (Public Law 92-500), also known as the Clean Water Act (CWA), stimulated substantial changes in wastewater treatment to achieve the objectives were not uniformly met.

Form the early 1970s to about 1980, wastewater treatment objective were based primarily on aesthetic and environmental concerns. The earlier objectives involving the reduction of biological oxygen demand (BOD), total suspended solids (TSS), and pathogenic organisms continued but at higher levels. Removal of nutrients, such as nitrogen and phosphorus, also began to be addressed, particularly in some of the inland streams and lakes, and estuaries and bays such as Chesapeake Bay and Long Island Sound. Major programs were undertaken by both states and federal agencies to achieve more effective and widespread treatment of wastewater to improve the quality of the surface waters. These programs were based, in part, on 1) an increased understanding of the environmental effects caused by wastewater discharges; 2) a greater appreciation of the adverse long-term effects caused by the discharge of the some of the specific constituents found in wastewater; and 3) the development of national concern for the protection of the environment. As a result of these programs, significant improvements have been made in the quality of the surface waters.

Since 1980, the water-quality improvement objectives of the 1970s have continued, but the emphasis has shifted to the definition and removal of constituents that may cause long-term health effects and environmental impacts. Health and environmental concerns are discussed in more detail in the following section. Consequently, while the early treatment objectives remain valid today, the required degree of treatment has increased significantly, and additional treatment objectives and goals have been added. Therefore, treatment objectives must go hand in hand with the water quality objectives or standards established by the federal, state, and regional regulatory authorities. Important federal regulations that have brought about changes in the planning and design of wastewater treatment facilities in the United States are summarized in Tab. 17.1. It is interesting to note that the clean air acts of 1970 and 1990 have had a significant impact on industrial and municipal wastewater programs, primarily through the implementation of treatment facilities for the control of emissions.

Tab. 17.1 Federal Regulations in United States

Regulation	Description
Clean Water Act (CWA) (Federal Water Pollution Control Act Amendments of 1972)	Establish the National Pollution Elimination System (NPDES), a permitting program based on uniform technological minimum standards for each discharger
Water Quality Act of 1987 (WQA) (Amendment of the CWA)	Strengthens federal water quality regulations by providing changes in permitting and adds substantial penalties for permit violations. Amends solids control program by emphasizing identification and regulation of toxic pollutants in sewage sludge
40 FR Part 503(1993) (Sewage Sludge Regulations)	Regulates the use disposal of biosolids from wastewater treatment plants. Limitations are established for items such as contaminants (mainly metals), pathogen content, and vector attraction
National Combined Sewer Overflow (CSO) Policy (1994)	Coordinates planning, selection, design, and implementation of CSO management practices and controls to meet requirements of CWA. Nine minimum controls and development of long-term CSO control plans are required to be implemented immediately
Clean Air Act of 1970 and 1990 Amendments	Establish limitations for specific air pollutants and institutes prevention of significant deterioration in air quality. Maximum achievable control technology is required for any of 189 listed chemicals form "major sources," i.e., plants emitting at least 60 kg/d
Total Maximum Daily Load (TMDL)(2000) Section 303(d) of the CWA	Requires states to develop prioritized lists of polluted or threatened water bodies and to establish the maximum amount of pollutant (TMDL) that a water body can receive and still meet water quality standards

U.S. Environmental Protection Agency (U.S.EPA) published its definition of minimum standard for secondary treatment. This definition, originally issued in 1973, was amended 1985 to allow additional flexibility in applying the percent removal requirements of pollutants to treatment facilities serving separates sewer systems. The definition of secondary treatment includes three major effluent parameters: 5-day BOD, TSS, and pH. The substitution of 5-day carbonaceous BOD ($CBOD_5$) for BOD_5 may be made at the option of the permitting authority. These standards provide the basis for design and operation of most treatment plants. Special interpretation of the definition of secondary treatment are permitted for publicly owned treatment works. 1) served by combined sewer systems, 2) using waste stabilization ponds are trickling filters, 3) receiving industrial flow, or 4) receiving less concentrated influent wastewater from separate sewers. The secondary treatment regulations were amended further in 1989 to clarify the percent removal requirements during dry period for treatment facilities served by combined sewers.

In 1987, Congress enacted the Water Act of 1987(WQA), the first may revision of the Clean Water Act. Important provisions of the WQA were: 1) strengthening federal water quality regulations by providing changes in permitting and added substantial penalties for permit violations, 2) significantly amending the CWA's form sludge control program by emphasizing the identification and regulation of toxic pollutants in sludge, 3) providing funding for state and U.S. EPA studies for defining non-point and toxic sources of pollution, 4) establishing new deadlines for compliant including priorities and permit requirements for stormwater, and 5) a phase-out of construction grants program as a method of financing publicly owned treatment won(POTW).

Recent regulations that affect wastewater facilities design include those for the treatment, disposal, and beneficial use of biosolids (40 CFR Part 503). In the biosolids regulation promulgated in 1993, national standards were set for pathogen and heavy metal content and for the handling and use of biosolids. The standards are designed to protect human health and the environment where biosoids are applied beneficially to land. The rule also promotes the development of a "clean sludge"(U.S.EPA, 1999).

The total maximum daily load (TMDL) program was promulgated in 2000 but is not scheduled to be in effect until 2002. The TMDL rule is designed to protect ambient water quality. A TDML represents the maximum amount of a pollutant that a water body can receive and still meet water quality standards. A TDML is the sum of 1) individual waste-load allocations for point sources, 2) load allocations for nonpoint sources, 3) natural background levels, and 4) a margin of safety (U.S.EPA, 2000). To implement the rule, a comprehensive watershed-based quality management program must be undertaken to find and control nonpoint sources in addition to conventional point source discharges. With implementation of the TDML rule, the focus on water quality shifts from technology-based controls to preservation of ambient water quality. The end result is an integrated planning approach that transcends jurisdictional boundaries and forces different sectors, such as agriculture, water and wastewater utilities, and urban runoff mangers, to cooperate. Implementation of the TMDL rule will vary depending on specific water quality objectives established for each watershed and, in some cases, will require the installation of advanced levels of treatment.

Vocabulary

floatable	漂浮的	secondary treatment	二级处理
aesthetic	美学的,审美的	combined	组合的
bay	海湾	revision	修订
estuary	河口	promulgate	公布,宣布
biosolid	生物固体		

Reading Material B

Health and Environmental Concerns in Wastewater Management

As research into the characteristics of wastewater has become more extensive, and as the techniques for analyzing specific constituents and their potential health and environmental effects have become more comprehensive, the body of scientific knowledge has expanded significantly. Many of the new treatment methods being developed are designed to deal with health and environmental concerns associated with findings of recent research. However, the advancement in treatment technology effectiveness has not kept pace with the enhanced constituent detection capability. Pollutants can be detected at lower concentrations than can be attainted by available treatment technology. Therefore, careful assessment of health and environment effects and community concerns about these effects becomes increasingly important in wastewater management. The need to establish a dialogue with the community is important to assure that health and environmental issues are being addressed.

Water quality issues arise when increasing amounts of treated wastewater are discharged to water bodies that are eventually used as water supplies. The waters of the Mississippi River and many rivers in the eastern United States are used for municipal and industrial water supplies and as repositories for the resulting treated wastewater. In southern California, a semiarid region, increasing amounts of reclaimed wastewater are being used or are planned to be used for groundwater recharge to augment existing potable water supplies. Significant questions remain about the testing and levels of treatment necessary to protect human health where the commingling of highly treated wastewater with drinking water sources results in indirect potable reuse. Some professionals object in principle to the indirect reuse of treated wastewater for potable purposes; others express concern that current techniques are inadequate for detecting all microbial and chemical contaminants of health significance (Crook et al., 1999). Among the latter concerns are 1) the lack of sufficient information regarding the health risks posed by some microbial pathogens and chemical constituents in wastewater, 2) the nature of unknown or unidentified chemical constituents and potential pathogens, and 3) the effectiveness of treatment processes for their removal. Defining risks to public health based on sound science is an ongoing challenge.

Because new and more sensitive methods for detecting chemicals are available and methods have been developed that better determine biological effects, constituents that were undetected previously are now of concern. Examples of such chemical constituents found in both surface and ground waters include: methyl tertiary butyl ether (MTBE), a highly soluble gasoline additive, medically active substances including endocrine disruptors, pesticides industrial chemicals, and phenolic compounds commonly found in nonionic surfactants. Endocrine-disrupting chemicals are a special health concern as they can mimic hormones produces in vertebrate animals by causing exaggerated response, or they can block the effects of a hormone on the body (Trussell 2000). These chemicals can cause problems with development, behavior, and reproduction in a variety of species. Increase in testicular, prostate, and breast cancers have been blamed on endocrine-disruptive chemicals (Roefer et al., 2000). Although treatment of these chemicals is not currently a mission of municipal wastewater treatment, wastewater treatment facilities may have to be designed to deal with these chemicals in the future.

Other health concerns relates to : 1) the release of volatile organic compounds (VOCs) and toxic air contaminants (TACs) from collections and treatment facilities, 2) chlorine disinfection, and 3) disinfection byproducts (DBPs). Odors are one of the most serious environment concerns to the public. New techniques for odor measurement are used to quantify the development and movement of odors that may emanate from wastewater facilities, and special efforts are being made to design facilities that minimize the development of odors, contain them effectively, and provide proper treatment for their destruction.

May industrial wastes contain VOCs that may be flammable, toxic, and odorous, and may be contributors to photochemical smog and tropospheric ozone. Provisions of 1) minimizing VOC releases at the source, 2) containing wastewater and their VOC emissions (i.e., by adding enclosures), treating wastewater foe VOC removal, and collecting and treating vapor emissions from wastewater. Many VOCs, classified as TACs, are discharged to the ambient atmosphere and transported to downwind receptors. Some air management districts are enforcing regulations based on excess cancer risks foe lifetime exposures to chemicals such as benzene, trichloroethylene, chloroform, and methylene chloride (Card and Corsi, 1992).

Effluents containing chlorine residuals are toxic to aquatic life, and increasingly, provisions to eliminate chlorine residuals are being instituted. Other important health issues relate to the reduction of disinfection byproducts (DBPs) that are potential carcinogens and are formed when chlorine reacts with organic matter. To achieve higher and more consistent microorganism inactivation levels, improved performance of disinfection systems must be addressed. In many communities, the issues of safety in the transporting, storing, and handling of chlorine are also being examined.

Vocabulary

assessment	评估	volatile	挥发性的
semiarid	半干旱的	photochemical smog	光化学烟雾
reclaimed wastewater	回用废水	tropospheric	对流层效应
constituent	成分	trichloroethylene	三氯乙烯
gasoline	汽油	methylene	亚甲基
phenolic	酚	foe	反对者,危害物

Unit 18 Wastewater Collection

The "Shambles" is a street or area in many medieval English cities, like London and York. During the eighteenth and nineteenth centuries, Shambles were commercialized areas, with meat packing as a major industry. The butchers of the Shambles would throw all of their waste into drainage ditches. The condition of the street was so bad that it contributed its name to the English language originally as a synonym for butchery or a bloody battlefield.

In old cities, drainage ditches like those at the Shambles were constructed for the sole purpose of moving stormwater out of the cities. In fact, discarding human excrement into these ditches was illegal in London. Eventually, the ditches were covered over and became what we now know as storm sewers. As water supplies developed and the use of the indoor water closet increased, the need for transporting domestic wastewater, called sanitary waste, became obvious. In the United States, sanitary wastes were first discharged into the storm sewers, which then carried both sanitary wastes and stormwater and were known as sanitary sewers. Eventually a new system of underground pipes, known as sanitary sewers, was constructed for removing the sanitary wastes. Cities and parts of cities built in the twentieth century almost all built separate sewers for sanitary waste and stormwater.

Estimating Wastewater Quantities

Domestic wastewater (sewage) comes from various sources within the home, including the washing machine, dishwasher, shower, sinks, and of course the toilet. The toilet, or water closet (WC), as it is still known in Europe, has become a standard fixture of modern urban society. As important as this invention is, however, there is some dispute as to its inventor. Some authors credit John Bramah with its invention in 1778; others recognize it as the brainchild of Sir John Harrinton in 1596. The latter argument is strengthened by Sir John's original description of the device, although there is no record of his donating his name to the invention. The first recorded is found in a 1735 regulation at Harvard University that decreed, "No Freshman shall go to the Fellows' John."

The term sewage is used here to mean only domestic wastewater. Domestic wastewater flows vary with season, the day of the week, and the hour of the day. Typically, average sewage flows are in the range of 378.5 per day per person, but especially in smaller communities that average can range widely.

Sewers also commonly carry industrial wastewater. The quantity of industrial wastes may usually be established by water use records, or the flows may be measured in manholes that serve only a specific industry, using a small flow meter. Industrial flows also often vary considerably throughout the day, the day of the week, and the season.

In addition to sewage and industrial wastewater, sewers carry groundwater and surface water that seeps into the pipes. Since sewer pipes can and often do have holes in them (due to faulty construction, cracking by roots, or other causes), groundwater can seep into the sewer pipe if the pipe is lower than the top of the groundwater table. This flow into sewers is called infiltration. Infiltration is least for new, well-constructed sewers, but can be as high as 500 m^3(km·day). For older systems, 700 m^3(km·day) is commonly estimated infiltration. Infiltration flow is detrimental since the extra volume of water must go through the sewers and the wastewater treatment plant. It should be reduced as much as possible by maintaining and repairing sewers and keeping sewerage easements clear of large trees whose roots can severely damage the sewers.

Inflow is stormwater collected unintentionally by the sanitary sewers. A common source of inflow is perforated manhole cover placed in a depression, so that stormwater flows into the manhole. Sewers laid next to creeks and drainage ways that rise up higher than the manhole elevation, or where the manhole is broken, are also a major source. Illegal connections to sanitary sewers, such as roof drains, can substantially increase the wet weather flow over the dry weather flow. The ratio of dry weather flow to wet weather flow is usually between 1:1.2 and 1:4.

For these reasons, the sizing of sewers is often difficult, since not all of the expected flows can be estimated and their variability is unknown. The more important the sewer and the more difficult to replace it, the more important it is to make sure that it is sufficiently large to be able to handle all the expected flows for the foreseeable future.

System Layout

Sewers collect wastewater from residences and industrial establishments. A system of sewers installed for the purpose of collecting wastewater is known as a sewerage system. Sewers also almost always operate as open channels or gravity flow conduits. Pressure sewers are used in a few places, but these are expensive to maintain and are useful only when there are severe restrictions on water use or when the terrain is such that gravity flow conduits cannot be efficiently maintained.

A typical system for a residential area is shown in Fig. 18.1. Building connections are usually made with clay or plastic pipe, 0.15 m in diameter, to the collecting sewers that run under the street. Collecting sewers are sized to carry the maximum anticipated peak flows without surcharging (filling up) and are ordinarily made of plastic, clay, cement, concrete, or cast iron pipe. They discharge into intercepting sewers, or interceptors, that collect from large areas and discharge finally into the wastewater treatment plant.

Collecting and intercepting sewers must be constructed with adequate slope for adequate flow velocity during periods of low flow, but not so step a slope as to promote excessively high velocities

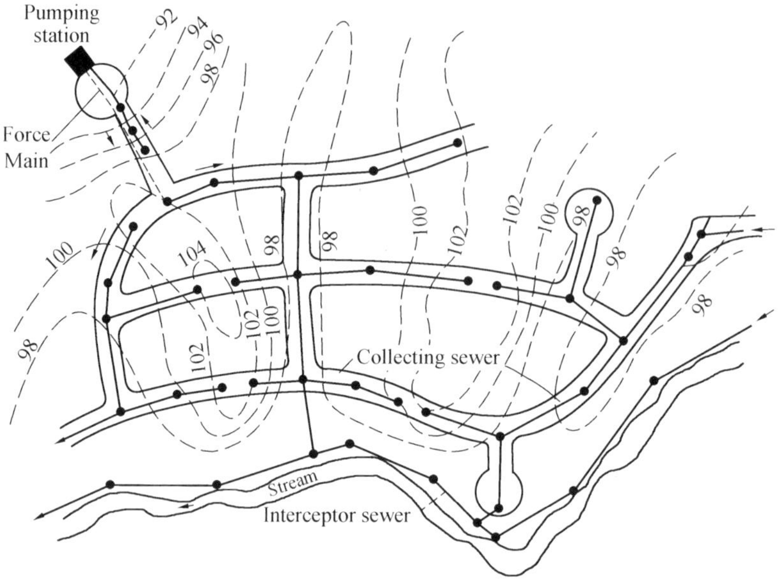

Fig. 18.1 Typical wastewater collection system layout

when flows are at their maximum. In addition, sewers must have manholes, usually every 120 to 180 m to facilitate cleaning and repair. Manholes are necessary whenever the sewer changes slope, size, or direction.

Gravity flow may be impossible, or uneconomical, in some locations so that the wastewater must be pumped. This requires the installation of pumping stations at various throughout the system. The pumping station collects wastewater from a collecting sewer and pumps it to a higher elevation by means of a force main. The end of a force main is always into a manhole.

A power outage would render the pumps inoperable, and eventually the sewage would back up into homes. As you can imagine, this would be highly undesirable; therefore, a good system layout minimizes pumping stations and/or provides auxiliary power.

Conclusion

Sewers have a part of civilized settlements for thousands of years, and in the modern Unite States we have become accustomed to and even complacent about the sewers that serve our communities. They never seem to fail, and there never seems to be a problem with them. Most important, we can dump whatever we want to down the drain, and it just disappears.

Of course, it doesn't just disappear. It flows through the sewer and ends up in a wastewater treatment plant. The stuff we often thoughtlessly dump down the drain can in fact cause serious problems in wastewater treatment and may even cause health problems in future drinking water supplies. Therefore, we must be cognizant of what we flush down the drain and recognize that it does not just disappear.

Vocabulary

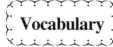

shambles	混乱的地方，废墟	pumping station	泵站
medieval	中世纪的	layout	规划，设计
excrement	排泄物，大便	velocity	速度
ditch	沟渠	become accustomed to	习惯于
illegal	违法的，不合规定的	cognizant	认知
sewage	下水道，污水	disappear	消失
infiltration	渗滤，渗透		

Reading Material A

Past and Present of U. S. Wastewater Treatment

Prior to 1948, the principal responsibility for controlling water pollution was assumed by the states and by various local and regional agencies. The first institutions to deal with water pollution problems were created soon after the "sanitary awakening" of the 1850s, when waterborne diseases reached epidemic proportions.

The federal role in water pollution control began with the Public Health Service Act of 1912. This act established the Streams Investigation Station at Cincinnati to carry out water pollution research. The Oil Pollution Act was passed in 1924 to prevent oily discharge on coastal waters. During the 1930s and 1940s, there was a continuing debate over whether the federal government should take a greater role in controlling water pollution. The debate led to the limited expansion of federal powers expressed in Water Pollution Control Act of 1948. The Federal Water Pollution Control Act (FWPCA) of 1956 was the cornerstone of early federal efforts to reduce pollution. Key elements of the act included a new program of subsidies for municipal treatment, plant construction, and an expanded basis for federal legal action against polluters. Increased funding for state water pollution control efforts and new support for research and training activities were also provided. Each of these programs was continued in the any amendments to the Federal Water Pollution Control Act in the 1960s and 1970s.

From the early 1970s to about 1980s, wastewater treatment objectives were based primarily on aesthetic and environmental concern, the earlier objectives of BOD, suspended solid, and pathogenic organisms reduction continued but at higher levels. Removal of nutrients (e. g., nitrogen, phosphorous) also began to be addressed for inland streams and lake. A major effort was undertaken by both state and federal agencies to achieve more effective and widespread treatment to improve the quality of the surface waters. This effort resulted in part from 1) an increased understanding of the environmental effects caused by wastewater discharges; 2) a developing knowledge of the adverse long-term effects caused by the discharges of some of the specific constituents found in wastewater; 3) the development of national concern for environmental protection. The result of these efforts was a significant improvement in the quality of the surface waters.

It was during this period that the concept of "Zero Discharge" was espoused. The goal was to treat wastewaters to the greatest possible extent, often without regard to the assimilative capacities of receiving waters, with the hope of eliminating rather than just controlling pollutant discharge to the environment. The result of this philosophy was that treatment systems became highly mechanistic processes contained in massive concrete and steel structures. Highly skilled personnel were required to efficiently operate systems and cost of construction, operation, and maintenance were high.

Since 1980, because of increased scientific knowledge and an expanded information base, wastewater treatment has begun to focus on the health concerns related to toxic and potentially toxic chemicals. The water-quality improvement objectives of the 1970s have continued, but the emphasis has shifted to definition and removal of compounds that may cause long-term health effects. As a consequence, while the early treatment objectives remain valid today, the required degree of treatment has increased significantly, and additional treatment objectives and goals have been added. At the same time, treatment requirements became somewhat relaxed in those cases where high levels of treatment were of no benefit to man or the environment. Both of these changes were prompted by a realization that man will inevitably have impact on the environment and that the goal of this treatment should be to minimize the effects.

Over the last forty years, the number of treatment plants serving municipalities and communities has nearly tripled. Implementation of the federal Clean Water Act brought about substantial changes in water pollution control to achieve "fishable and swimmable" waters. Over 15,000 facilities are in operation, according to the U. S. Environmental Protection Agency's recent needs survey. An analysis of the data on the sizes of treatment plants shows that approximately 81 percent of all publicly owned treatment works with treatment needs are smaller than 43.8 L/s. However over 66 percent of the total wastewater volume treated is handled by facilities having capacities greater than 438 L/s.

Because of the high potential for environmental degradation from the discharge from the large plants as a result of the amount of contaminants delivered to these facilities, levels of treatment are accordingly high. Therefore, most of these larger facilities are highly mechanistic and require a great deal of operator skill and control. Further, they are very costly to construct, operate and maintain. However, the smaller plants are often non-mechanistic systems which require minimal attention by an operator. This results in poorer levels of treatment at lower cost, as compared to the larger plants. The trade-offs at the smaller facilities come from the fact of smaller amounts of pollution being delivered to these plants, thereby reducing their potential impact on the environment.

Vocabulary

prior to	在…之前	amendment	修正，改正
sanitary	卫生的，清洁的	Zero Discharge	零排放
cornerstone	基石	degradation	降解
Water Pollution Control Act	水污染防治法案		

Reading Material B

Wastewater Collection System

Sanitary Sewers

Sanitary sewers carry domestic sewage, liquid commercial and industrial wastes, and undesirable contributions from infiltration and stormwater. Because sewage discharges from homes, buildings, and factories can occur simultaneously, sanitary sewers must be designed to handle the peak rate of flow. The Harmon formula is one of several formulae, based on population, that are used to estimate the ratio of peak to average flowrate.

In some municipalities, sink disposal units may be installed in sinks to macerate household organic(food)wastes so that the wastes can be flushed to the sanitary sewer rather than collected with the municipal garbage. The convenience to the homeowners and the possibility of less frequent garbage pickup are offsetted by the cost of the unit and the increased cost for treating the additional SS and organic load discharged to the sewers and municipal treatment facilities. Where sink disposal unites are used, increases in the SS and BOD of the sewage have been estimated at 30 and 60 percent, respectively.

Storm Sewers

Storm sewers receive stormwater runoff from roads, roofs, lawns, and other surfaces. Various methods for estimating the rates or volumes of runoff are available. All start with a storm of specified frequency (from 2 to 100 years) and information on an actual (or representation) storm, or with curves relating rainfall intensity to storm duration. Typical storm frequencies used for design are 2 to 5 years for residential areas, 10 to 25 years for high-valve and commercial districts, and 50 to 100 years for major drainage systems.

Storm system may be partial systems with relatively small pipes, providing only road drainage and not connected to building drains. Because these sewers are not connected to buildings, they can surcharge(overflow)without flooding basements, causing only minor inconvenience. On the other hand, complete storm systems that provide both road drainage and storm connections to buildings must have lager pipes designed with less likelihood of surcharging if backup of stormwater around basement footings is to be avoided.

Buildings with basements are protected against high groundwater levels by the installation of perforated or open-jointed drainage pipes laid in a gravel trench around the basement footings. The uncontaminated water collected by these foundation drains may then discharge to 1) a storm or combined sewer if sewer backup is unlikely (a common practice in larger cities); 2) a sanitary sewer or basement sump (typical of smaller municipalities); or 3) a separate foundation drain collector (limited to occasional installations where a third street sewer with a free outlet can be installed).

As has been mentioned previously, runoff can contain high concentrations of pollutants. Increasing public awareness of the pollution caused by stormwater forces of sanitary and storm sewers are common in the older sections of most large municipalities. Because these sewers cause a backup of untreated sanitary sewage into basements. Consequently, such systems must be designed to accommodate large storms (recurring at 25-to-50 year average intervals) without surcharging, while still receiving sanitary flow. Since the volume of sanitary sewage, in comparison with the peak amount of stormwater flow, is insignificant, it can normally be neglected in design.

Pollution from Combined Sewer Overflows

During dry weather all flow in combined sewers goes to the treatment plant. During rainfall, the plant may be able accept 1.5 to 3 times the dry weather flow (DWF), but the excess stormwater, now combined with municipal wastewater, must be discharge to the receiving waters without proper treatment. In a city where all sanitary sewage is collected for secondary treatment (90 percent removal of BOD and SS), the pollution (in terms of BOD and SS) from untreated combined sewer overflows can exceed that from the treated sanitary sewage. Most solutions to the problem are expensive, and none are completely satisfactory. Two basic approaches have been used: either some degree of separation or storage of some proportion of the excess flow for subsequent treatment.

Complete separation of stormwater from municipal wastewater has generally been considered the best long-term solution. Unfortunately for many cities, this solution is economically impractical. Washington, D.C., is one of the few large cities to have adopted the method. Partial separation, a less costly alternative, is another approach. Toronto is implementing this scheme in some older section of the city at about half the cost of full separation. The new sewers provide road drainage and create reserve storage capacity in the existing combined sewer system to less than half the number of previous occurrences. It can also be used to allow redevelopment of low-density housing with higher density apartment development, which of course creates increased sanitary sewage flows in account of the much higher population.

Vocabulary

flowrate	流量	concentration	浓度
simultaneously	同时地	impractical	不合实际的
macerate	浸泡,泡软	Toronto	多伦多(加拿大)
residential area	住宅区		

Unit 19 Reactors Used for the Treatment of Wastewater

Wastewater treatment involving physical unit operations and chemical and biological unit processes is carried out in vessels or tanks commonly known as "reactors". The type of reactors that are available and their applications are introduced in this section.

Types of Reactors

The principal type of reactors used for the treatment of wastewater, illustrated on the Fig. 19.1, are 1) the batch reactor, 2) the complete-mix reactor (also knows as the continuous-flow stirred-tank reactor (CFSTR) in the chemical engineering literature), 3) the plug-flow reactor (also known as a tubular-flow reactor), 4) complete-mix reactors in series, 5) the packed-bed reactor, and 6) the fluidized-bed reactor. Brief descriptions of these reactors are presented below.

Batch Reactor In the batch reactor (see Fig. 19.1 (a)), flow is neither entering nor leaving the reactor (i.e., flow enters, is treated, and then is discharged, and the cycle repeats). The liquid contents of reactor are mixing completely. Batch reactors are often used to blend chemicals or to dilute concentrated chemicals.

Complete-mix Reactor In the complete-mix reactor (see Fig. 19.1 (b)), it is assumed that complete mixing occurs instantaneously and uniformly throughout the reactor as fluid particles enter the reactor. Fluid particles leave the reactor in proportion to their statistical population. Complete mixing can be accomplished in round or square reactors if the contents of the reactor are uniformly and continuously redistributed. The actual time required to achieve completely mixed conditions will depend on the reactor geometry and the power input.

Plug-flow Reactor Fluid particles pass through the reactor with little or no longitudinal mixing and exit from the reactor in the same sequence in which they entered. The particles retain their identity and remain in the reactor for a time equal to the theoretical detention time. This type of flow is approximated in long open tanks with a high length-to-width ratio in which longitudinal dispersion is minimal or absent (see Fig. 19.1(c)) or closed tubular reactors (e.g., pipelines, see Fig. 19.1(d)).

Complete-mix Reactors in Series The series of complete-mix reactors (see Fig. 19.1(e)) is used to model the flow regime that exists between the ideal hydraulic flow patterns corresponding to the complete-mix and plug-flow reactors. If the series is composed of one reactor, the complete-mix regime prevails. If the series consists of an infinite number of reactors in series, the plug-flow regime prevails.

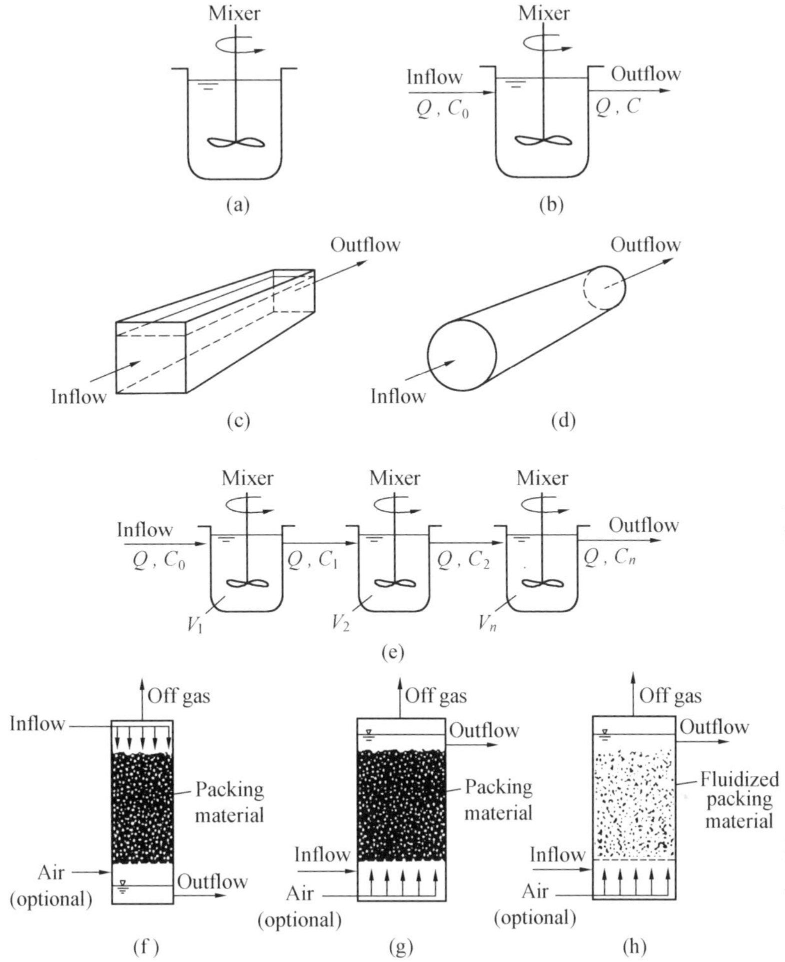

Fig. 19.1 Different types of reactors used for wastewater treatment
(a) batch reactor, (b) complete-mix reactor, (c) plug-flow open reactor, (d) plug-flow closed reactor also known as a tubular reactor, (e) complete-mix reactor in series, (f) packed-bed reactor, (g) packed-bed upflow reactor, and (h) expanded-bed upflow reactor

Packed-bed Reactors The packed-bed reactor is filled with some type of packing material, such as rock, slag, ceramic, or now more commonly, plastic. With respect to flow, the packed-bed reactor can be operated in either the downflow or upflow mode. Dosing can be continuous or intermittent (e.g., trickling filter). The packing material in packed-bed reactors can be continuous (see Fig. 19.1 (f)) or arranged in multiple stages, with flow from one stage to another. A packed-bed upflow anaerobic (without oxygen) reactor is shown in Fig. 19.1(g).

Fluidized-bed Reactor The fluidized-bed reactor is similar to the packed-bed reactor in many respects, but the packing material is expanded by the upward movement of fluid (air or water) through the bed (see Fig. 19.1(h)). The expanded porosity of the fluidized-bed packing material can be varied by controlling the flow-rate of the fluid.

Application of Reactors

The principal applications of reactor types used for wastewater treatment are reported in Tab. 19.1. Operational factors that must be considered in the selection of the type of reactor or reactors to be used in the treatment process include 1) the nature of the wastewater to be treated, 2) the nature of the reaction (i.e., homogeneous or heterogeneous), 3) the reaction kinetics governing the treatment process, 4) the process performance requirements, and 5) local environmental conditions. In practice, the construction costs and operation and maintenance costs also affect reactor selection. Because the relative importance of these factors varies with each application, each factor should be considered separately when the type of reactor is to be selected.

Tab. 19.1 Principal applications of reactor types used for wastewater treatment

Type of reactor	Application in wastewater treatment
Batch	Activated-sludge biological treatment in a sequence batch reactor, mixing of concentrated solutions into working solutions
Complete-mix	Aerated lagoons, aerobic sludge digestion
Complete-mix with recycle	Activated-sludge biological treatment
Plug-flow	Chlorine contact basin, natural treatment systems
Plug-flow with recycle	Activated-sludge biological treatment, aquatic treatment systems
Complete-mix reactors in series	Lagoon treatment systems, used to simulate non-ideal flow in plug-flow reactors
Packed-bed	Non-submerged and submerged tricking-filter biological treatment units, depth filtration, natural treatment systems, air stripping
Fluidized-bed	Fluidized-bed reactor for aerobic and anaerobic biological treatment, upflow sludge blanket reactors, air stripping

Vocabulary

unit operation	单元操作	longitudinal	纵向的
unit process	单元过程	homogeneous	同类的
batch	序,批	heterogeneous	异类的
complete-mix	完全混合	activated-sludge	活性污泥
fluidized-bed	流化床	digestion	消化
instantaneously	瞬间	submerged	淹没的

Reading Material A

Treatment Processes Involving Mass Transfer

Common operations and processes in wastewater treatment involving mass transfer are identified in Tab. 19.2. The most important mass transfer operations in wastewater treatment involving 1) the transfer of material across gas-liquid interfaces as in aeration and in the removal of unwanted constituents from wastewater by air stripping, and 2) the removal of unwanted constituents from wastewater by adsorption onto solid surfaces such as activated carbon and ion exchange. To introduce the concepts involved in mass transfer, the basic principle of mass transfer is reviewed followed by a consideration of gas-liquid and liquid-solid mass transfer operations.

Tab. 19.2 Principal applications of mass transfer operations and process in wastewater treatment

Type of reactor	Phase equilibrium	Application
Absorption	gas→liquid	Addition of gases to water
Adsorption	gas→solid liquid→solid	Removal of organics with activated carbon Removal of organics with activated carbon, dechlorination
Desorption	solid→liquid solid→gas	Sediment scrubbing Reactivation of spent activated
Dying(evaporation)	liquid→gas	Drying of sludge
Gas stripping	liquid→gas	Removal of gases
Ion exchange	liquid→solid	Selective removal of chemical constituents, demineralization

Gas-liquid Mass Transfer

Over the past 50 years a number of mass transfer theories have been proposed to explain the mechanism of gas transfer across gas-liquid interfaces. The simple and most commonly used is the two-film theory proposed by Lewis and Whitman (1924). The penetration model proposed by Higbie (1935) and the surface-renewal model proposed by Danckwerts (1951) are more theoretical and take into account more of the physical phenomena involved. The two-film theory remains popular because, in more than 95 percent of the situations encountered, the results obtained are essentially the same as those obtained with the more complex theories. Even in the 5 percent where there is some disagreement between the two-film theory and other theories, it is not clear which approach is more correct.

The two-film theory is based on a physical model in which two films exist at the gas-liquid interface, as shown in Fig. 19.2. Two conditions are shown in Fig. 19.2: a) "absorption," in which a gas is transferred from the gas phase to the liquid phase, and b) "desorption," in which a gas in transferred out of the liquid phase into the phase. The two films, one liquid and one gas, provide the resistance to the passage of gas molecules between the bulk-liquid and the bulk-gaseous phases. It is very important to note that in the application of the two-film theory it is assumed that the concentration and partial pressure in both the bulk-liquid and bulk-gaseous phase are uniform (i. e., mixed completely).

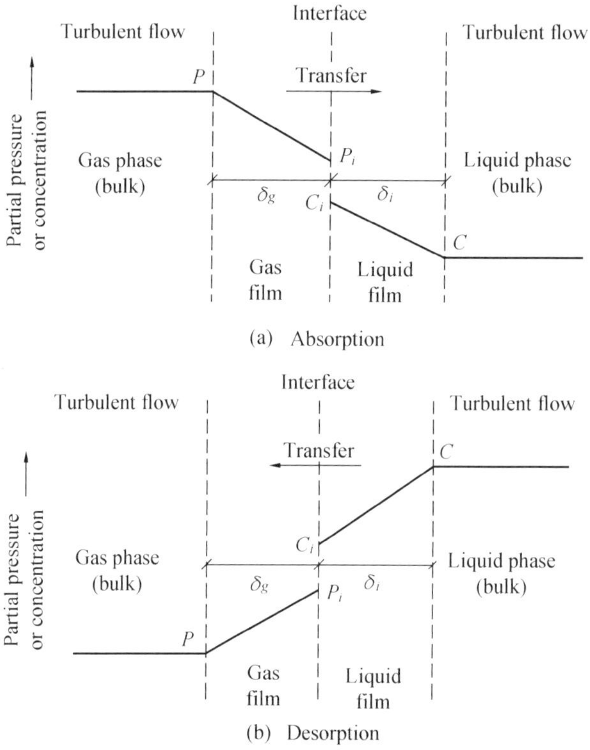

Fig 19.2 Definition sketch for the two-film theory of gas transfer

Liquid-solid Mass Transfer

In the discussion of gas-liquid mass transfer, it was found that mass could be transferred from either phase to the other phase. In liquid-solid mass transfer operations, constituents from the liquid phase are transferred (adsorbed) to a solid phase. Adsorption and ion exchange mass transfer processes are the example of liquid-solid mass transfer.

The process of accumulating substances that are in solution on a suitable interface is termed adsorption. The adsorbate is the substance that is being removed from liquid phase at the interface. The adsorbent is the solid, liquid, or gas phase onto which the adsorbate accumulates. The adsorption process takes place in three steps: macro-transport, micro-transport, and sorption. Granular or powered activated carbon (GAC or PAC) is used most commonly for the removal of selected constituents from wastewater. The accumulation of material is described by what is known as adsorption isotherm, which is used to define the mass of material adsorbed per unit mass of adsorbing material. A common adsorption isotherm is the Freundlich isotherm.

Ion exchange is a mass transfer process in which ions of a given species are displaced from an insoluble exchange material by ions of a different species in solution. In water treatment, ion exchange is used most commonly to soften hard water through the removal of multivalent cations. In wastewater treatment, the principal uses for 1) the removal of nitrogen and phosphorus, 2) the removal of heavy metals, and 3) the removal of total dissolved solid (i.e., demineralization) for reuse applications.

Vocabulary

transfer	传递,转移	dechlorination	脱氯
interface	界面	adsorbate	被吸附物
equilibrium	平衡	phenomena	现象
adsorption	吸附	adsorption isotherm	吸附等温线
absorption	吸收	cation	阳离子
desorption	解吸,解吸作用		

Reading Material B

Wastewater Treatment

Prior to about 1940, most municipal wastewater was generated from domestic sources. After 1940, as industrial development in the United States grew significantly, increasing amounts of industrial wastewater have been and continue to be discharged to municipal collection systems. The amounts of heavy metals and synthesizes organic compounds generated by industrial activities have increased, and some 10,000 new organic compounds are added each year. Many of these compounds are now found in the wastewater from most municipalities and communities.

As technological changes take place in manufacturing, changes also occur in the compounds discharged and the resulting wastewater characteristics. Numerous compounds generated from industrial processes are difficult and costly to treat by conventional wastewater treatment processes. Therefore, effective industrial pretreatment becomes an essential part of an overall water quality management program.

Wastewater collected from municipalities and communities must ultimately be returned to receiving waters or to the land or reused. The complex question facing the design engineer and public health officials is: What levels of treatment must be achieved in a given application—beyond those prescribed by discharge permits—to ensure protection of public health and the environment? The answer to this question requires detailed analyses of local conditions and needs, application of scientific knowledge and engineering judgment based on past experience, and consideration of federal, state, and local regulations. In some cases, a detailed rise assessment may be required. An overview of wastewater treatment is provided in this section.

Treatment Methods

Methods of treatment in which the application of physical forces predominate are known as unit operations. Methods of treatment in which removal of contaminants is brought about by chemical or biological reactions are known as unit processes. At the present time, unit operations and processes are grouped together to provide various levels of treatment known as preliminary, primary, advanced primary, secondary (without or with nutrient removal), and advanced (or tertiary) treatment. In preliminary treatment, gross solids such as large object, rags, and grit are removed that may damage equipment. In primary treatment, a physical operation, usually sedimentation, is used to remove the floating and settleable materials found in wastewater. For advanced primary treatment, chemicals are added to enhanced the removal of suspended solids and, to a lesser extent, dissolved solids. In secondary treatment, biological and chemical processes are used to remove most of the organic matter. In advanced treatment, additional combinations of unit operations and processes are used to remove residual suspended solids and other constituents that are not reduced significantly by conventional secondary treatment.

About 20 years ago, biological nutrient removal (BNR)—for the removal of nitrogen and phosphorus—was viewed as an innovative process for advanced wastewater treatment. Because of the extensive research into the mechanisms of BNR, the advantages of its use, and the number of BNR systems that have been placed into operation, nutrient removal, for all practical purposes, has become a part of conventional wastewater treatment. When compared to chemical treatment methods, BNR uses less chemical, reduces the production of waste solids, and has lower energy consumption. Because of the importance of BNR in wastewater treatment, BNR is integrated into the discussion of theory, application, and design of biological treatment system.

Land treatment processes, commonly termed "natural systems," combine physical, chemical, and biological treatment mechanisms and produce water with quality similar to or better than that from advanced wastewater treatment. Natural systems are not covered in this text as they are used mainly with small treatment systems.

Current Status

Up until the late 1980s, conventional secondary treatment was the most common method of treatment for the removal of BOD and TSS. In the United States, nutrient removal was used in special circumstances, such as in the Great Lakes area, Florida, and the Chesapeake Bay, where sensitive nutrient-related water quality conditions where identified. Because of nutrient enrichment that has led to eutrophication and water quality degradation (due in part to point source discharge), nutrient removal processes have evolved and now are used extensively in other areas as well.

As a result of implementation of the Federal Water Pollution Control Act Amendments, significant data have been obtained on the numbers and types of wastewater facilities used and needed in accomplishing the goals of the program. Surveys are conducted by U.S.EPA to track these data, and the results of the 1996 Needs Assessment Survey (U.S. EPA, 1997) are reported. These data are useful in forming an overall view of the current status of wastewater treatment in the United States.

The municipal wastewater treatment enterprise is composed of over 16,000 plants that are used to treat a total flow of about 1 400 cubic meters per second (m^3/s). Approximately 92 percent of the total existing flow is handled by plants having a capacity of 0.044 m^3/s and larger. Nearly one-half of the present design capacity is situated in plants providing greater than secondary treatment. Thus, the basic material presented in this text is directed toward the design of plants larger than 0.044 m^3/s with the consideration that many new designs will provide treatment greater than secondary.

In the last 10 years, many plants have been designed using BNR. Effluent filtration has also been installed where the removal of residual suspended solids is required. Filtration is especially effective in improving the effectiveness of disinfection, especially for ultraviolet (UV) disinfection systems, because 1) the removal of larger particles of suspended solids that harbor bacteria enhances the reduction in coliform bacteria and 2) the reduction of turbidity improves the transmittance of UV light. Effluent reuse systems, except for many that are used for agricultural irrigation, almost always employ filtration.

New Directions and Concerns

New directions and concerns in wastewater are evident in various specific areas of wastewater treatment, the changing nature of the wastewater to be treated, emerging health and environmental concerns, the problem of industrial wastes, and the impact of new regulations. All of which have been discussed previously, are among the most important. Further, other important concerns include: 1) aging infrastructure, 2) new methods of process analysis and control, 3) treatment plant performance and reliability, 4) wastewater disinfection, 5) combined sewer overflows, 6) impacts of stormwater and sanitary overflows and nonpoint sources of pollution, 7) separate treatment of return flows, 8) odor control and the control of VOC emissions, and 9) retrofitting and upgrading wastewater treatment plants.

Future Trends in Wastewater Treatment

In the U.S. EPA Needs Assessment Survey, the total treatment plant design capacity is projected to increase by about 15 percent over the next 20 to 30 years. During this period, the U.S. EPA estimates that approximately 2,300 new plants may have to be built, most of which will be providing a level of treatment greater than secondary treatment is expected to increase by 40 percent in the future (U.S.EPA, 1997). Thus, it is clear that the future trends in wastewater treatment plant design will be for facilities providing higher levels of treatment.

Some of the innovative treatment methods being utilized in new and upgraded treatment facilities include vortex separators, high rate clarification, membrane bioreactors, press-driven membrane filtration (ultrafiltration and reverse osmosis), and ultraviolet radiation (low-pressure, low-and high-intensity UV lamps, and medium-pressure, high-intensity UV lamps). Some of the new technologies, especially those developed in Europe, are more compact and are particularly well suited for plants where available space for expansion is limited.

In recent years, numerous proprietary wastewater treatment processes have been developed that offer potential savings in construction and operation. This trend will likely continue, particularly where alternative treatment systems are evaluated or facilities are privatized. Privatization is generally defined as a public-private partnership in which the private partner arranges the financing, design, building, and operation of the treatment facilities. In some cases, the private partner may own the facilities. The reasons for privatization, however, go well beyond the possibility of installing proprietary processes. In the United States, the need for private financing appears to be the principal rationale for privatization; the need to preserve local control appears to be the leading pragmatic rationale against privatization.

Vocabulary

synthesize	合成,综合	vortex	旋涡,涡流
conventional	传统的,常规的	membrane bioreactor	膜生物反应器
eutrophication	富营养化	ultrafiltration	超滤
reverse osmosis	反渗透	ultraviolet radiation	紫外辐射

Unit 20 Flotation

Flotation is a unit operation used to separate solid or liquid particles form a liquid phase. Separation is brought about by introducing fine gas (usually air) bubbles into the liquid phase. The bubbles attach to the particulate matter, and the buoyant force of the combined particle and gas bubbles is great enough to cause the particle to rise to the surface. Particle that have a higher density than the liquid can thus be made to rise. The rising of particles with lower density than the liquid can also be facilitated (e.g., oil suspension in water).

In wastewater treatment, flotation is used principally to remove suspended matter and to concentrate biosolids. The principal advantages of flotation over sedimentation are that very small or light particles that settle slowly can be removed more completely in a shorter time. Once the particles have been floated to the surface, they can be collected by operation.

Description

The present practice of flotation as applied to wastewater treatment is confined to the use of air as the flotation agent. Air bubbles are added or caused to form by 1) injection of air while the liquid is under pressure, followed by release of the pressure (dissolved-air flotation), and 2) aeration at atmospheric (dispersed-air flotation). In these systems, the degree of removal can be enhanced through the use of the various chemical additives. In municipal wastewater treatment, dissolved-air flotation is frequently used, especially for thickening of waste biosolids.

Dissolved-air Flotation

In dissolved-air flotation (DAF) systems, air is dissolved in the wastewater under a pressure of several atmospheres, followed by release of the atmospheric level (see Fig. 20.1). In small pressure systems, the entire flow may be pressurized by means of a pump to 275 to 350 kPa with compressed air added at the pump suction (see Fig. 20.1(a)). The entire flow is held in a retention tank under pressure for several minutes to allow time for the air to dissolve. It is then admitted through a pressure-reducing valve to the flotation tank where the air comes out of solution in very fine bubbles.

In the larger units, a portion of the DAF effluent (15 to 120 percent) is recycled, pressurized, and semi-saturated with air (Fig. 20.1(b)). The recycled flow is mixed with the unpressurized main stream just before admission to the flotation tank, with the result that the air comes out of solution in contact with particulate matter at the entrance to the tank. Pressure types of units have been mainly for the treatment of industrial wastes and for the concentration of solids.

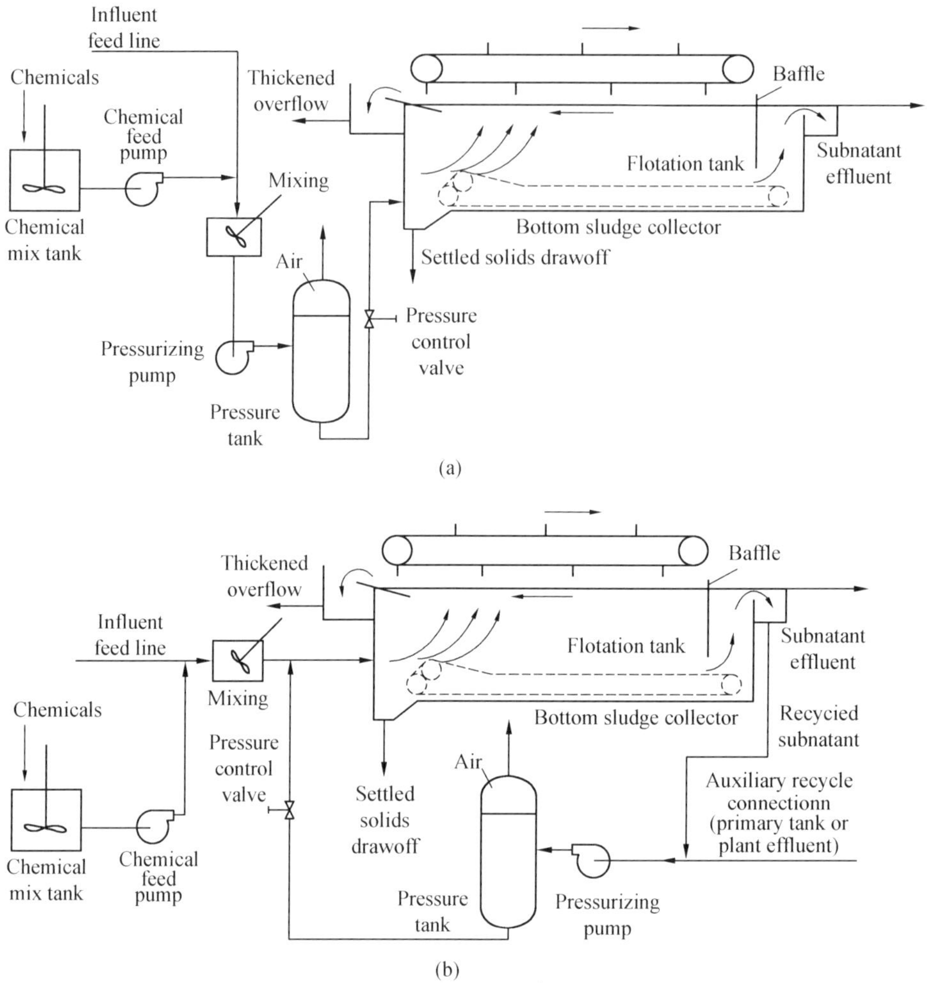

Fig. 20.1 Schematic of dissolved-air flotation systems: (a) without recycle in which the entire flow is passed through the pressurizing tank; (b) with recycle in which only the recycle flow is pressurized

Dispersed-air Flotation

Dispersed-air (sometimes referred to as induced-air) flotation is seldom used in municipal wastewater treatment, but it is used in industrial application for the removal of emulsified oil and suspended solids from high-volume waste or process waters. In dispersed-air flotation systems, air bubbles are formed by introducing the gas phase directly into the liquid phase through a revolving impeller. The spinning impeller acts a pump, forcing fluid through disperser opening and creating a vacuum in the standpipe (see Fig. 20.2). The vacuum pulls air (or gas) into the standpipe and thoroughly mixes it with the liquid. As the gas/liquid mixture travels through the disperser, a mixes

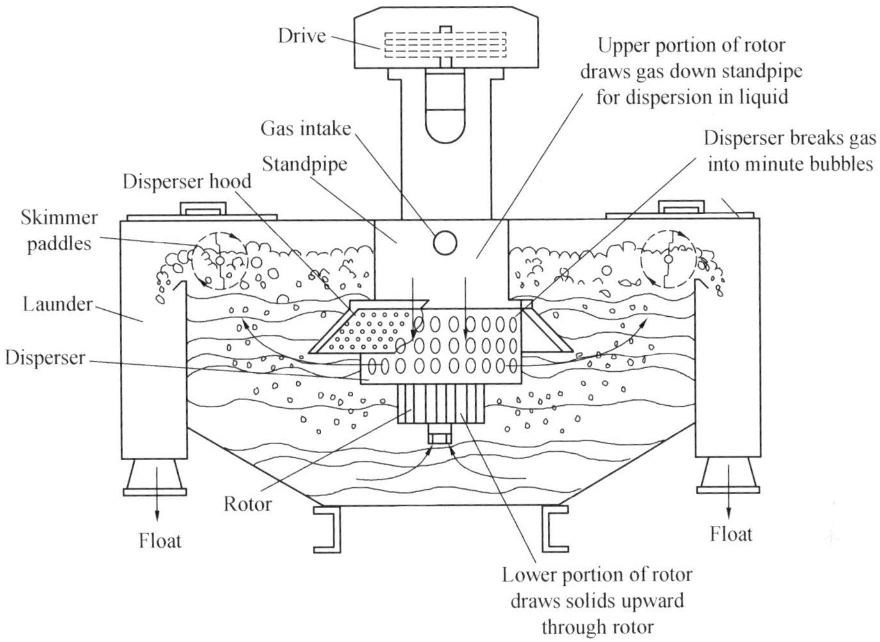

Fig. 20.2 Dispersed-air flotation

force is created that causes the gas to form very fine bubbles. The liquid moves through a series of cells before leaving the unit. Oil particles and suspended solids attach to the bubbles as they rise to the surface. The oil and suspended solids gather in dense froth at the surface and are removed by skimming paddles. The advantages of a dispersed-air flotation system are 1) compact size, 2) lower capital cost, and 3) capacity to remove relatively flee oil and suspended solids. The disadvantages of induced-air flotation include higher connected power requirements than the pressurized system, performance is dependent on strict hydraulic control, and less flocculation flexibility. The quantities of float skimmings are significantly higher than the pressurized unit: 3 to 7 percent of the incoming flow as compared to less than 1 percent for dissolved-air systems (Eckenfelder, 2000).

Chemical Additives

Chemicals are commonly used to aid the flotation process. These chemicals, for the most part, function to create a surface or a structure that can easily adsorb or entrap air bubbles. Inorganic chemicals, such as the aluminum and ferric salts and activated silica, can be used to bind the particulate matter together and in so doing, create a structure that can easily entrap air bubbles. Various organic polymers can be used to change the nature of either the air-liquid interface or the solid-liquid interface, or both. These compounds usually collect on the interface to bring about the desired changes.

Vocabulary

flotation	浮选,气浮	pressure-reducing valve	减压阀
buoyant	有浮力的,上涨的	unpressurized	非受压的
dissolved-air flotation	溶气气浮	vacuum	真空
suction	吸入,吸力	dispersed-air flotation	散气气浮

Reading Material A

Primary Sedimentation

The objective of treatment by sedimentation is to remove readily settleable solids and floating material and thus reduce the suspended solids content. Primary sedimentation is used as a preliminary step in the further processing of the wastewater. Efficiently designed and operated primary sedimentation tanks should remove from 50 to 70 percent of the suspended solids and from 25 to 40 percent of the BOD.

Sedimentation tanks have also been used as stormwater retention tanks, which are designed to provide a moderate detention period (10 to 30 min) for overflows from either combined sewers or storm sewers. The purpose of sedimentation is to remove a substantial portion of the organic solids that otherwise would be discharged directly to the receiving waters. Sedimentation tanks have also been used to provide detention periods sufficient for effective disinfection of such overflows.

Almost all treatment plants use mechanically cleaned sedimentation tanks of standardized circular or rectangular design. The selection of the type of sedimentation unit for a given application is governed by the size of the installation, by rules and regulations of local control authorities, by local site conditions, and by the experience and judgment of the engineer. Two or more tanks should be provided so that the process may remain in operation while one tank is out of service for maintenance and repair work. At large plants, the number of tanks is determined largely by size limitations.

Rectangular Tanks

Rectangular sedimentation tanks may use either chain-and-flight solids collectors or traveling-bridge-type collectors. A rectangular tank uses a chain-and flight-type collector. Equipment for settled solids removal generally consists of a pair of endless conveyor chains, manufactured of alloy steel, cast iron, or thermoplastic. Attached to the chains at approximately 3-m intervals are scraper flights made of wood or fiberglass, extending the full width of the tank or bay. The solids settling in the tank are scraped to solids hoppers in small tanks and to transverse troughs in large tanks. The transverse troughs are equipped with collecting mechanisms (cross collectors), usually either chain-and flight or screw-type collectors, which convey solids to one or more collection hoppers. In very long units (over 50 m), two collection mechanisms can be used to scrape solids to collection points near the middle of the tank length. Where possible, it is desirable to locate solids pumping facilities close to the collection hoppers.

Rectangular tanks may also be cleaned by a bridge-type mechanism that travels up and down the tank on rubber wheels or on rails supported on the sidewalls. One or more scraper blades are suspended from the bridge. Some of the bridge mechanisms are designed so that the scraper blades can be lifted clear of the solids blanket on the return travel.

Where cross collectors are not provided, multiple solids hoppers must be installed. Solids hoppers have operating difficulties, notably solids accumulation on the slopes and in the corners and arching over the solids draw off piping. Wastewater may also be drawn through the solids hopper, bypassing some of the accumulated solids, resulting in a "rathole" effect. A cross collector is more advisable, except possibly in small plants, because a more uniform and concentrated solids can be withdrawn and many of the problems associated with solids hoppers can be eliminated.

Because flow distribution in rectangular tanks is critical, one of the following inlet designs is used: 1) full-width inlet channels with inlet weirs, 2) inlet channels with submerged ports or orifices, 3) or inlet channels with wide gates and slotted baffles. Inlet weirs, while effective in spreading flow across the tank width, introduce a vertical velocity component into the solids hopper that may resuspend the solids particles. Inlet ports can provide good distribution across the tank width if the velocities are maintained in the 3 to 9 m/min range. Inlet baffles are effective in reducing the high initial velocities and distribute flow over the widest possible cross-sectional area. Where full-width baffles are used, they should extend from 150 mm below the surface to 300 mm below the entrance opening.

For installations of multiple rectangular tanks, below-grade pipe and equipment galleries can be constructed integrally with the tank structure and along the influent end. The galleries are used to house the sludge pumps and sludge drawoff piping. The galleries also provide access to the equipment for operation and maintenance. Galleries can also be connected to service tunnels for access to other plant units.

Scum is usually collected at the effluent end of rectangular tanks with the flights returning at the liquid surface. The scum is moved by the flights to a point where it is trapped by baffles before removal. Water sprays can also move the scum. The scum can be scraped manually up an inclined apron. Or it can be removed hydraulically or mechanically, and for scum removal a number of means have been developed. For small installations, the most common scum drawoff facility consists of a horizontal, slotted pipe that can be rotated by a lever or a screw. Except when drawing scum, the open slot is above the normal tank water level. When drawing scum, the pipe is rotated so that the open slot is submerged just below the water level, permitting the scum accumulation to flow into the pipe. Use of this equipment results in a relatively large volume of scum liquor.

Another method for removing scum is by a transverse rotating helical wiper attached to a shaft. Scum is removed from the water surface and moved over a short inclined apron for discharge to a cross-collecting scum trough. The scum may then be flushed to a scum ejector or hopper ahead of a scum pump. Another method of scum removal consists of a chain-and-flight type of collector that collects the scum at one side of the tank and scrapes it up a short incline for deposit in scum hoppers, whence it can be pumped to disposal units. Scum is also collected by special scum rakes in rectangular tanks that are equipped with the carriage or bridge type of sedimentation tank equipment. In installations where appreciable amounts of scum are collected, the scum hoppers are usually equipped with mixers to provide a homogeneous mixture prior to pumping. Scum is usually disposed of with the solids and biosolids produced at the plant; however, separate scum disposal is used at many plants.

Multiple rectangular tanks require less land area than multiple circular tanks and find application where site space is at a premium. Rectangular tanks also lend themselves to nesting with preaeration tanks and aeration tanks in activated-sludge plants, thus permitting common wall construction and reducing construction costs. They are also used generally where tank roofs or covers are required.

Circular Tanks

In circular tanks the flow pattern is radial (as opposed to horizontal in rectangular tanks). To achieve a radial flow pattern, the wastewater to be settled can be introduced in the center or around the periphery of the tank, as shown in Fig. 20.3. Both flow configurations have proved to be satisfactory generally, although the center-feed type is more commonly used, especially for primary treatment. In the center-feed design (see Fig. 20.3 (a)), the wastewater is transported to the center of the tank in a pipe suspended from the bridge, or encased in concrete beneath the tank floor. At the center of the tank, the wastewater enters a circular well designed to distribute the flow equally in all directions. The center well has a diameter typically between 15 and 20 percent of the total tank diameter and ranges from 1 to 2.5 m in depth and should have a tangential energy-dissipating inlet within the feedwell.

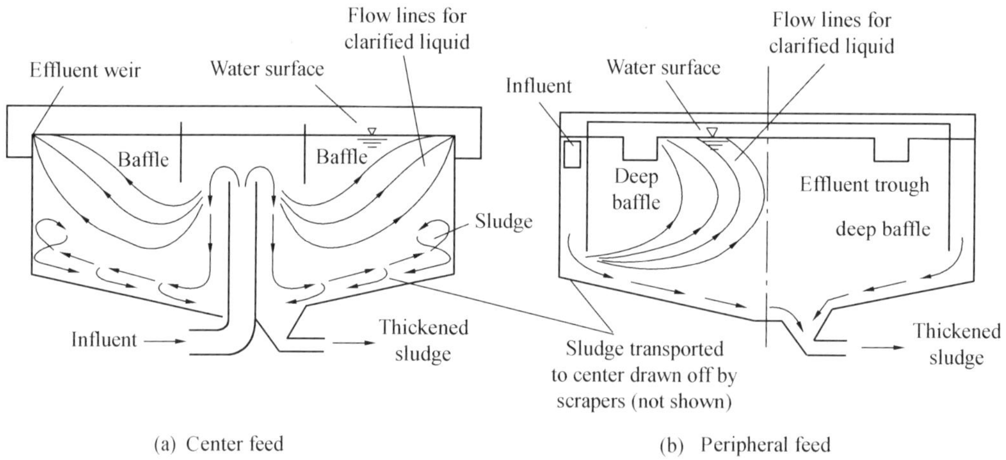

Fig. 20.3 Typical circular primary sedimentation tank

The energy-dissipating device functions to collect influent from the center column and discharge it tangentially into the upper 0.5 to 0.7 m of the feedwell. The discharge ports are sized to produce a velocity of < 0.75 m/s at maximum flow and 0.30 to 0.45 m/s at average flow. The feedwell should be sized so that the maximum downward velocity does not exceed 0.75 m/s. The depth of the feedwell should extend about 1 meter below the energy-dissipating inlet ports (Randall et al., 1992).

In the peripheral-feed design (see Fig. 20.3(b)), a suspended circular baffle forms an annular space into which the inlet wastewater is discharged in a tangential direction. The wastewater flows spirally around the tank and underneath the baffle, and the clarified liquid is skimmed off over weirs on both sides of a centrally locates weir trough. Grease and scum are confined to the surface of the annular space. Peripheral feed tanks are used generally for secondary clarification.

Circular tanks 3.6 to 9 m in diameter have the solids-removal equipment supported on beams spanning the tank. Tanks 10.5 m in diameter and larger have a central pier that supports the mechanism and is reached by a walkway or bridge.

The bottom of the tank is sloped at about 1 in 12 (vertical : horizontal) to form an inverted cone, and the solids are scraped to a relatively small hopper located near the center of the tank.

Multiple tanks are customarily arranged in grouped of two or four. The flow is divided among the tanks by a flow-split structure, commonly located between the tanks. Solids are usually withdrawn by sludge pumps for discharge to the solids processing and disposal units.

Vocabulary

sedimentation	沉淀,沉降	horizontal	水平的
rectangular	矩形的	diameter	直径
hopper	漏斗	tangential	切线的
scum	浮渣	apron	挡板，护板
aeration	通风,通氧		

Reading Material B

Grit Removal System

The main function of the grit removal process at any wastewater treatment plant is the removal of heavy, inorganic material such as sand and gravel from the wastewater flow, to prevent undue abrasion and wear on equipment, and to reduce the potential for deposition and accumulation of solids in downstream channels, tanks and pipes where the grit may obstruct flow and be difficult to remove.

Typically, sand and gravel particles are prevalent in grit. However, seeds, cinders, coffee grounds, etc. will also be found in grit solids. Excess amounts of grit can be present in wastewater conveyed by either separate or combined sewer systems, with far more being found in combined systems.

Depending on the type of grit removal process used, the removed grit is often further concentrated in a cyclone or classifier, and washed to remove any organic material which has been captured alone with the grit. The washed grit is more readily disposed of. The quantity and characteristics of the grit removed from wastewater can vary over a wide range depending on the type of collection system, characteristics of the drainage area, use of household garbage grinders, presence of industrial wastes, and the efficiency of the grit removal equipment. It is generally accepted that sand and other inorganic matter which is 100 mesh (0.15 mm) or coarser in size will cause most trouble and that at least 95 percent of this material should be removed before the sewage flow undergoes treatment. To this end, most grit removal facilities are designed to remove all grit larger than 100 mesh during normal flow and with a somewhat lesser efficiency at maximum flow (possibly grit coarser than 5 mesh under these conditions). The grit removal facilities should be designed to operate under widely varying flows with a minimum loss of head. Any grit removed should be thoroughly drained of free water and relatively free of organic matter.

A variety of grit removal devices or processes can be utilized including:

1. Aerated grit chamber;
2. Vortex type system;
3. Detritus tank system;
4. Horizontal flow (constant velocity) grit channel.

The Greater Moncton Sewerage Commission Plant uses two aerated grit chambers for grit removal. The aerated grit chambers usually provide a minimum prescribed detention time, and are designed to introduce compressed air along one side near the bottom of the tank to cause a "spiral roll" velocity. The rolling action is independent of flow, allowing the aerated grit chamber to operate effectively over a wide range of flows.

The main advantages of the aerated grit chamber are that the same efficiencies of removal can be accomplished over a wide range of incoming flow without any necessary adjustments, and this removal is accomplished with very low head loss. The major disadvantages are the power consumption required for the air blowers and additional labour for maintenance and control of aeration process.

A submerged screw conveyor mounted in the recess in each chamber moves the settled grit to the suction of a pump which pumps the grit to the grit dewatering process.

The grit slurry which is removed from the bottom of each chamber is pumped by the grit pumps to the degritting cyclones. The cyclones separate the grit material from the organics in the grit slurry by using the centrifugal forces developed in the cyclone to cause the heavier grit particles to concentrate along the sides and bottom, and the lighter organic matter to be removed from the centre through the top of each cyclone. The heavier grit solids are removed through the bottom and are discharged to the grit dewatering classifier. The lighter organic matter is returned back to the inlet of the grit removal process.

The grit dewatering classifier consists of an inclined screw type mechanism which washes and dewaters the grit by separating the heavy grit solids from the lighter organic solids and the liquid. The grit solids are removed from the liquid by the screw mechanism which discharges the grit through a discharge chute to the inlet of a screw conveyor.

The screw conveyor conveys the grit to a grit dumpster. The grit dumpster is emptied as required by "load-lugger" truck and the grit is removed for disposal.

Aerated Grit Chamber Air Blowers

The air blowers that supply the air to the aerated grit chambers and aerated channels have been sized to provide enough capacity for the ultimate design. A total of three (3) blowers have been installed. At the present time one duty unit is used along with a standby unit of the same capacity. A larger capacity unit is used for aeration during high flow conditions.

Grit Chamber Screw Conveyors

Grit removed in the aerated grit tanks is collected in a hopper at the bottom of each tank and is transferred to one end of the tank by the grit screw conveyor, to the suction of the grit removal pump.

Grit Pumps

Each screw conveyor transfers the removed grit to the suction of the grit removal pump. There is one grit pump associated with each grit tank (grit tank screw conveyor), however, the suctions to the grit pumps are cross connected to allow for flexibility in case of a grit pump outage. In future, there will be a grit pump for each additional grit tank which may be added. These additional pumps will also be cross connected, in as much as there will not be any standby grit pumps.

Each grit pump discharge to a cyclone degritter. The pump discharges are not cross connected.

Cyclone Degritters

The grit slurry removed from the bottom of the grit tanks is pumped by the grit pumps to degritting cyclones. The cyclones separate the grit from the waste stream by using the centrifugal forces created by the grit pump inlet pressure which enters the stationary cylindrical body of the cyclone tangentially. The cyclone has a liquid discharge located at the top of the unit and a solids discharge at the lower end. The action of the flow entering the cyclone creates a vortex or spiral flow pattern which forces the grit particles to walls of the unit where these are directed to the bottom outlet. The lighter, grit free wastewater (which may contain some organic solids), flows to the centre of the vortex and exits through the overflow or liquid discharge.

The grit stream from each cyclone is sent to a grit classifier for further rinsing and draining of the material prior to disposal. The grit free wastewater is returned to the inlet of the aerated grit tanks.

Grit Classifier

The classifier washes and dewaters the underflow from the cyclones. It is comprised of a grit settling tank, screw conveyor, an overflow weir (adjustable) and a flushing water supply.

Grit separation takes place in the settling tank and the conveyor moves the grit up an incline to the classifier discharge. Flushing water is added here when the classifier is operating to maintain a clear channel for drainage. Excess water overflows the weir and is returned to the grit tank inlet channel.

Grit Screw Conveyor

The grit solids are conveyed from the classifier discharge chute to the disposal bins.

Vocabulary

grit	砂,粗砂	cinder	煤渣
abrasion	磨损	vortex	旋流
accumulation	集聚	screw	螺旋
obstruct	阻隔，阻塞，阻碍物，障碍物	degritter	除砂机

Unit 21 Chemical Oxidation

Chemical oxidation in wastewater treatment typically involves the use of oxidizing agents such as ozone, hydrogen peroxide, permanganate, chloride dioxide, chlorine or (HOCl), and oxygen, to bring about change in the chemical composition of a compound or a group of compounds. Included in the following discussion is an introduction of the fundamental concepts involved in chemical oxidation, an overview of the uses of chemical oxidation in wastewater treatment, and a discussion of the use of chemical oxidation for the reduction of BOD and COD, the oxidation of ammonia, and oxidation of nonbiodegradable organic compounds. Advanced oxidation process (AOPs) in which the free hydroxyl radical is used as a strong oxidant to destroy specific organic constituents and compounds that cannot be oxidized by conventional oxidants such as ozone and chlorine, which deals which advanced treatment methods.

Fundamentals of Chemical Oxidation

The purpose of the following discussion is to introduce the basic concepts involved in chemical oxidation reactions. The topics to be discussed include 1) oxidation-reduction reactions, 2) half reaction potentials, 3) reaction potentials, 4) equilibrium constants for redox equations, and 5) rate of oxidation-reduction reactions.

Oxidation-reduction Reactions

Oxidation-reduction reactions (known as redox equations) take place between an oxidizing agent and a reducing agent. In oxidation-reduction reactions both electrons are exchanged as are the oxidation states of the constituents involved in the reaction. While an oxidizing agent causes the oxidation to occur, it is reduced in the process. Similarly, a reducing agent that causes a reduction to occur is oxidized in the process. For example, consider the following reduction:

$$Cu^{2+} + Zn \leftrightarrow Cu + Zn^{2+} \qquad (21.1)$$

In the above reaction copper (Cu) changes from a +2 to zero oxidation state and the zinc (Zn) changes from a zero to a +2 state. Because of the electron gain or loss, oxidation-reduction reactions can be separated into two half reactions. The oxidation half reaction involves the loss of electrons while the reduction half reaction involves the gain of electrons. The two half reactions that comprise Eq. (21.1) are as follows:

$$Zn - 2e^- \leftrightarrow Zn^{2+} \qquad \text{(oxidation)} \ (21.2)$$

$$Cu^{2+} + 2e^- \leftrightarrow Cu \qquad \text{(reduction)} \ (21.3)$$

Referring to the above equations, there is a two-electron change.

Half-reaction Potentials

Because of the almost infinite number of possible reactions, there are no summary tables of equilibrium constants for oxidation-reduction reactions. What is done instead is the chemical and thermodynamic characteristics of the half reactions, such as those given by Eqs. (21.2) and (21.3), are determined and tabulated so that any combination of reactions can be studied. Of the many properties that can be used to characterize oxidation reduction reactions, the electrical potential (i.e., voltage) or emf of the half reaction is used most commonly. Thus, every half reaction involving an oxidation or reduction has a standard potential E^0 associated with it. The potentials for the reactions given by Eqs. (21.2) and (21.3) are as follows:

$$Cu^{2+} + 2e^- \leftrightarrow Cu \quad E^0 = 0.34\ V \tag{21.4}$$

$$Zn - 2e^- \leftrightarrow Zn^{2+} \quad E^0 = -0.763\ V \tag{21.5}$$

The half-reaction potential is a measure of the tendency of a reaction to proceed to the right. Half reactions with large positive potential, E^0, tend to proceed to the right as written. Conversely, half reactions with large negative potential, E^0, tend to proceed to the left.

Reaction Potentials

The half-reaction potentials, discussed above, can be used to predict whether a reaction comprised of two half reactions will proceed as written. The tendency of a reaction to proceed is obtained by determining the for the E^0_{reaction} entire reaction as given by the following expression.

$$E^0_{\text{reaction}} = E^0_{\text{reduction}} - E^0_{\text{oxidation}} \tag{21.6}$$

where E^0_{reaction} = potential of the overall reaction;

E^0_{reaction} = potential of the reduction half reaction;

E^0_{reaction} = potential of the oxidation half reaction.

For example, for the reaction between copper and zinc [see Eq. (21.1)] the E^0_{reaction} of the reaction is determined as follows:

$$E^0_{\text{reaction}} = E^0_{Cu^{2+},Cu} - E^0_{Zn^{2+},Zn} \tag{21.7}$$

$$E^0_{\text{reaction}} = 0.34 - (-0.763) = +1.103\ V \tag{21.8}$$

The positive value for the E^0_{reaction} is taken as an indication that the reaction will proceed as written. The magnitude of the value, as will be illustrated subsequently, can be taken as a measure of the extent to which the reaction as written will proceed. For example, if Eq. (21.1) had been written as follows:

$$Cu + Zn^{2+} \leftrightarrow Cu^{2+} + Zn \tag{21.9}$$

The corresponding E^0_{reaction} for this reaction is

$$E^0_{\text{reaction}} = E^0_{Zn^{2+},Zn} - E^0_{Cu^{2+},Cu} \tag{21.10}$$

$$E^0_{\text{reaction}} = (-0.763) - 0.34 = -1.103\ V \tag{21.11}$$

Because the E^0_{reaction} for the reaction is negative, the reaction will proceed in the opposite direction from what is written.

Rate of Oxidation-reduction Reactions

As noted previously, the half-reaction potentials can be used to predict whether a reaction will proceed as written. Unfortunately, the reaction potential provides no information about the rate at which the rate will proceed. Chemical oxidation reactions often require the presence of one or more catalysts for the reaction to proceed or to increase the rate of reaction. Transition metal cations, enzymes, pH adjustment, and a variety of proprietary substances have been used as catalysts.

Vocabulary

oxidizing agent	氧化剂	equilibrium constant	平衡常数
hydrogen peroxide	过氧化氢	redox	氧化还原作用
nonbiodegradable	生物不能降解的	reduction	还原
advanced oxidation process	高级氧化过程	catalyst	催化剂
oxidation-reduction reaction	氧化还原反应		

Reading Material A

Types of Mixers Used for Continuous Mixing in Wastewater Treatment

Continuous mixing operations are used in biological treatment processes such as the activated-sludge process to maintain the mixed liquor suspended solids uniformly mixed state. In biological treatment systems the mixing device is also used to provide the oxygen needed for the process. Thus, the aeration equipment must be able to provide the oxygen needed for the process and the energy needed to maintain mixed conditions within the reactor. Both mechanical aerators and dissolved aeration devices are used. Diffused air is often used to fulfill both the mixing and oxygen requirements. Alternatively, mechanical turbine-aerator mixers may be used. Horizontal, submersible propeller mixers are often used to maintain channel velocities in oxidation ditches, mix the contents of anoxic reactors, and aid in the destratification of reclaimed water storage reservoirs.

Pneumatic Mixing

In pneumatic mixing, a gas (usually air or oxygen) is injected into the bottom of mixing or activated-sludge tanks, and the turbulence caused by the rising gas bubbles serves to mix the fluid contents of the tank. In aeration, soft bubbles are formed with an average diameter of 5 mm while the air flow is about 10 percent of the liquid flow. The velocity gradients due to bubble formation range from a $G_{avg} < 200\ s^{-1}$ to $G_{max} = 8\,200\ s^{-1}$ (Masschelein, 1992).

Where air flocculation is employed, the air supply system should be adjustable so that the flocculation energy level can be varied throughout the tank.

Mechanical Aerators and Mixers

The principal types of mechanical aerators used for continuous mixing are high-speed surface aerators and slow-speed surface aerators. These devices deals with aeration. Typical power requirements for mixing with mechanical aerators range from 20 to 40 $kW/10^3\ m^3$, depending on the type of mixer and the geometry of the tank, lagoon or basin.

New Developments in Mixing Technology

New analytical tools that are now being applied to the analysis of and design of mixing devices include 1) computational fluid dynamics (CFD), 2) digital particle image velocimetry (DPIV), 3) laser doppler anemometry (LDA), and 4) laser-induced fluorescence (LIF). Computational fluid dynamics is used to model the fluid flow patterns in mixing devices and for scale-up analysis. In respect to fluid flow, both two-and three-dimensional models are now available. Digital particle image velocimetry is used to understand fluid movement in mixing devices. The movement of neutrally buoyant fluorescent particles is photographed using laser beam illumination. Laser doppler anemometry is used to study turbulence and to obtain data on the mean velocity at a given location in the mixing chamber. To evaluate the mean velocity, two laser beams are focused so that the beams intersect. As a particle passes through the intersection of the beams, light is reflected. The wavelength of the reflected light is a function of the particle velocity. Laser-induced fluorescence is used to measure the mixedness of solutions. Dyes such as rhodimine and other materials will fluoresce when struck by laser light of a given wavelength. The scattering of light is measured to assess the degree of mixedness. This technique is being used to study the diffusion and mixing of a substance by assessing the coefficient of variation of the mixed solution and to evaluate blending times (Chemineer, Inc., 2000).

Vocabulary

device	装置	pneumatic	空气的,气体的
turbine	涡轮	velocimetry	速度测量学
propeller	推进器	fluorescence	荧光,荧光性
oxidation ditche	氧化沟	doppler	多普勒效应
destratification	混合,使不成层		

Reading Material B

Chemical Precipitation for Phosphorus Removal

The removal phosphorus from wastewater involves the incorporation of phosphate into TSS and the subsequent removal of those solids. Phosphorus can be incorporated into either biological solids (e.g., microorganisms) or chemical precipitates. The removal of phosphorus in chemical precipitates is introduced in this section. The topics to be considered include 1) the chemistry of phosphate precipitation, 2) strategies for phosphorus removal, 3) phosphorus removal using metal salts and polymers, and 4) phosphorus removal using lime.

Chemistry of Phosphate Precipitation

The chemical precipitation of phosphorus is brought about by the addition of the salts of multivalent metal ions that from precipitates of sparingly soluble phosphates. The multivalent metal ions used most commonly are calcium [Ca(II)], aluminum [Al(III)], and iron [Fe(III)]. Polymers have been used effectively in conjunction with alum and lime as flocculant aids. Because the chemistry of phosphate precipitation with calcium is quite different than with aluminum and iron, the two different types of precipitation are considered separately in the following discussion.

Phosphate Precipitation with Calcium

Calcium is usually added in the form of lime $Ca(OH)_2$. From the equations presented previously, it will be notes that when lime is added to water it reacts with the natural bicarbonate alkalinity to precipitate $CaCO_3$. As the pH value of the wastewater increases beyond about 10, excess calcium ions will then react with the phosphate, as shown in Eq. (21.12), to precipitate hydroxylapatite $Ca_{10}(PO_4)_6(OH)_2$.

$$10Ca^{2+} + 6PO_4^{3-} + 2OH^- \leftrightarrow Ca_{10}(PO_4)_6(OH)_2 \quad (21.12)$$

Because of the reaction of lime with the alkalinity of the wastewater, the quantity of lime required will, in general, be independent of the amount of phosphate present and will depend primarily on the alkalinity of wastewater. The quantity of lime required to precipitate the phosphorus in wastewater is typically about 1.4 to 1.5 times the total alkalinity expressed as $CaCO_3$. Because a high pH value is required to precipitate phosphate, coprecipitation is usually not feasible. When lime is added to raw wastewater or to secondary effluent, pH adjustment is usually required before subsequent treatment or disposal. Recarbonation with carbon (CO) is used to lower the pH value.

Phosphate Precipitation with Aluminum and Iron

The basic reactions involved in the precipitation of phosphorus with aluminum and iron are as follows.

Phosphate precipitation with aluminum:

$$Al^{3+} + H_nPO_4^{3-n} \leftrightarrow AlPO_4 + nH^+ \quad (21.13)$$

Phosphate precipitation with iron:

$$Fe^{3+} + H_nPO_4^{3-n} \leftrightarrow FePO_4 + nH^+ \quad (21.14)$$

In the case of alum and iron, 1 mole will precipitate 1 mole of phosphate; however, these reactions are deceptively simple and must be considered in light of the many competing reactions and their associated equilibrium constants, and the effects of alkalinity, pH, trace elements, and ligands found in wastewater. Because of the many competing reactions, Eqs. (21.13) and (21.14) cannot be used to estimate the required chemical dosages directly. Therefore, dosages are generally established on the basis of bench-scale tests and occasionally by full-scale test, especially if polymers are used. For example, for equimolar initial concentrations of A1(Ⅲ), Fe(Ⅲ), and phosphate, the total concentration of soluble phosphate in equilibrium with both insoluble $A1PO_4$ and $FePO_4$. The solid lines trace the concentration of residual soluble phosphate after precipitation. Pure metal phosphates are precipitated within the shaded area, and mixed complex polynuclear species are formed outside toward higher and lower pH values.

Vocabulary

biological phosphorus removal	生物除磷	mole	摩尔
chemical precipitate	化学沉淀	ligand	配位,配位体
polymer	聚合物	equimolar	当量分子
coprecipitation	一起沉淀	polynuclear	多环的,多核的
reaction	反应		

Unit 22 Wastewater Biological Treatment Processes

The objective of wastewater treatment is to reduce the concentrations of specific pollutants to the level at which the discharge of the effluent will not adversely affect the environment or pose a health threat. Moreover, reduction of these constituents need only be to some required level.

For any given wastewater in a specific location, the degree and type of treatment are variables that require engineering decisions. Often the degree of treatment depends on the assimilative capacity of the receiving water. DO sag curves can indicate how much BOD must be removed from wastewater so that the DO of the receiving water is not depressed too far. The amount of BOD that must be removed is an effluent standard and dictates in large part the type of wastewater treatment required.

To facilitate the discussion of wastewater, assume a "typical wastewater" and assume further that the effluent from this wastewater treatment must meet the following effluent standards:

BOD $\leqslant$ 15 mg/L,

SS $\leqslant$ 15 mg/L,

P $\leqslant$ 1 mg/L.

Additional effluent standards could have been established, but for illustrative purposes we consider only these three. The treatment system selected to achieve these effluent standards includes

1. Primary treatment: physical processes that nonhomogenizable solids and homogenize the remaining effluent.

2. Secondary treatment: biological process that remove most of the biochemical demand for oxygen.

3. Tertiary treatment: physical, biological, and chemical processes to remove nutrients like phosphorus and inorganic pollutants, to deodorize and decolorize effluent water, and to carry out further oxidation.

Primary Treatment

The most objectionable aspect of discharging raw sewage into watercourses is the floating material. Thus screens were the first form of wastewater treatment used by communities, and they are used today as the first step in treatment plants. Typical screens consist of a series of steel bars that might be about 2.5 cm apart. A screen in a modern treatment plant removes materials that might damage equipment or hinder further treatment. In some older treatment plants screens are cleaned by hand, but mechanical cleaning equipment is used in almost all new plants. The cleaning rakes are activated when screens get sufficiently clogged to raise the water level in front of the bars.

In many plants, the second treatment step is a comminutor, a circular grinder designed to grind the solids coming through the screen into pieces about 0.3 cm or less in diameter.

The third treatment step is the removal of grit or sand from the wastewater. Grit and sand can damage equipment like pumps and flow meters and must be removed. The most common grit chamber is a wide place in the channel where the flow is slowed enough to allow the dense grit to settle out. Sand is about 2.5 times denser than most organic solids and thus settles much faster. The objective of a grit chamber is to remove sand and grit without removing organic material. Organic material must be treated further in the plant, but the separated sand may be used as fill without additional treatment.

Most wastewater treatment plants have a settling tank after the grit chamber, to settle out as much solid material as possible. Accordingly, the retention time is long and turbulence is kept to a minimum.

The solids settle to the bottom of the tank and are removed through a pipe, while the clarified liquid escapes over a V-notch weir that distributes the liquid discharge equally all the way around a tank. Settling tanks are also called sedimentation tanks or clarifiers. The settling tank that immediately follows screening and grit removal is called the primary clarifier. The solids that drop to the bottom of a primary clarifier are removed as raw sludge.

Raw sludge generally has a powerfully unpleasant odor, is full of pathogenic organisms, and is wet, three characteristics that make its disposal difficult. It must be stabilized to retard further decomposition and dewatered for ease of disposal.

The objective of primary treatment is the removal of solids, although some BOD is removed as a consequence of the removal of decomposable solids.

A substantial fraction of the solids has been removed, as well as some BOD and a little P, as a consequence of the removal of raw sludge. After primary treatment the wastewater may move on to secondary treatment.

Secondary Treatment

Water leaving the primary clarifier has not lost much of the solid organic matter but still contains high-energy molecules that decompose by microbial action, creating BOD. The demand for oxygen must be reduced (energy wasted) or else the discharge may create unacceptable condition in the receiving waters. The objective of secondary treatment is to remove BOD, whereas the objective of primary treatment is to remove solids.

The trickling filter consists of a filter bed of fist-sized rocks or corrugated plastic blocks over which the waste is trickled. The name is something of a misnomer since no filtration takes place. A very active biological growth forms on the rocks, and these organisms obtain their food from the waste stream dripping through the rock bed. Air either is forced through the rocks or circulates automatically because of the difference between the air temperature in the bed and ambient temperatures. Trickling filters use a rotating arm that moves under its own power, like a lawn sprinkler, distributing the waste evenly over the entire bed. Often the flow is recirculated and a higher degree of treatment attained.

Trickling filtration was a well-established treatment system at the beginning of the twentieth century. In 1914, a pilot plant was built for a different system that bubbled air through free-floating aerobic microorganisms, a process which became known as the activated sludge system. The activated sludge process differs from trickling filtration in that the microorganisms are suspended in the liquid.

An activated sludge system includes a tank full of waste liquid from the primary clarifier and a mass of microorganisms. Air bubbled into this aeration tank provides the necessary oxygen for survival of the aerobic organisms. The microorganisms come in contact with dissolved organic matter in the wastewater, adsorb this material, and ultimately decompose the organic material to CO_2, H_2O, some stable compounds, and more microorganisms.

When most of the organic material, that is, food for the microorganisms, has been used up, the microorganisms are separated from the liquid in a settling tank, sometimes called a secondary or clarifier. The microorganisms remaining in the settling tank have no food available, become hungry, and are thus activated-hence the term activated sludge. The clarified liquid escapes over a weir and may be discharged into the receiving water. The settle microorganisms, now called return activated sludge, are pumped back to the head of the aeration tank, where they find more food in the organic compounds in the liquid entering the aeration tank from the primary clarifier, and the process starts over again. Activated sludge treatment is a continuous process, with continuous sludge pumping and clean-water discharge.

Activated sludge treatment produces more microorganisms than necessary and if the microorganisms are not removed, their concentration will soon increase and clog the system with solids. Some of the microorganisms must therefore be wasted and the disposal of such waste activated sludge is one of the most difficult aspects of wastewater treatment.

Activated sludge systems are designed on the basis of loading, or the amount of organic matter, or food, added relative to the microorganisms available. The food-to-microorganism (F/M) ratio is a major design parameter. Both F and M are difficult to measure accurately, but may be approximated by influent BOD and SS in the aeration tank, respectively. The combination of liquid and microorganisms undergoing aeration is known as mixed liquor, and the SS in the aeration tank are mixed liquid suspended solids (MLSS). The ratio of influent BOD to MLSS, the F/M ratio, is the loading on the system, calculated as pounds (or kg) of BOD per day per pound or kg of MLSS.

Relatively small F/M, or little food for many microorganisms, and a long aeration period (long retention time in the tank) result in a high degree of treatment because the microorganisms can make maximum use of available food. Systems with these features are called extended aeration systems and are widely used to treat isolated wastewater source, like small developments or resort hotels. Extended aeration systems create little excess biomass and little excess activated sludge to dispose of.

The success of the activated sludge system also depends on the separation of the microorganisms in the final clarifier. When the microorganisms do not settle out as anticipated, the sludge is said to be a bulking sludge. Bulking is often characterized by a biomass composed almost totally of filamentous organisms that form a kind of lattice structure within the sludge floes which prevents settling. A trend toward poor settling may be the forerunner of a badly upset and ineffective system. The settle ability of activated sludge is most often described by the sludge volume index (SVI), which is reasoned by allowing the sludge to settle for minutes in a 1-L cylinder. If the SVI is 100 or lower, the sludge solids settle rapidly and the sludge returned to the final clarifier can be expected at a high solids concentration. SVI is about 200, however, indicate bulking sludge and can lead to poor treatment.

Tertiary Treatment

The effluent from secondary treatment meets the previously established effluent standards for BOD and SS. Only phosphorus content remains high. The removal of inorganic compounds, including inorganic phosphorus and nitrogen compounds, requires advanced or tertiary wastewater treatment.

Primary and secondary (biological) treatments are a part of conventional wastewater treatment plants. However, secondary treatment plant effluents are still significantly polluted. Some BOD and suspended solids remain, and neither primary nor secondary treatment is effective in removing phosphorus and other nutrients or toxic substances. A popular advanced treatment for BOD removal is the polishing pond, or oxidation pond, commonly a large lagoon into which the secondary effluent flows. Such ponds have a long retention time, often measured in weeks.

BOD may also be removed by activated carbon adsorption, which has the added advantage of removing some is a completely enclosed tube, which dirty water is pumped into at the bottom and clear water exits at the top. Microscopic crevices in the carbon catch and hold colloidal and smaller particles. As the carbon column becomes saturated, the pollutants must be removed from the carbon and the carbon reactivated, usually by heating it in the absence of oxygen. Reactivated or regenerated carbon is somewhat less efficient than using virgin carbon, some of which must always be added to ensure effective performance.

Vocabulary

assimilative	同化的,吸收的	microorganism	微生物
DO sag curves	氧垂曲线	parameter	参数
nonhomogenizable	非均匀化的	sludge volume index	污泥指数
comminutor	粉碎,粉碎机	tertiary wastewater treatment	废水三级处理
trickling filtration	滴滤		

Reading Material A

Modification to Existing Processes

Biotower

The development of plastic media as the packing for trickling filter has been followed by the development of tall tower filters without the need for support. These tower bioreactors, because of their high surface area, can handle a high BOD loading, and in some cases forced air has been supplied to the base of the tower to increase the rate of degradation. The tower are used mainly to reduce BOD in high-strength wastes and often added to a conventional system ahead of the aeration tank to smooth out variations in BOD and to increase plant capacity.

Rotating Biological Contactor

The availability of plastic media has also seen another development—the rotating biological contactor. Here a drum of honeycomb plastic or closely spaced discs is slowly rotated (1-2 r/min) with the base (40 percent) of the drum in the settled sewage or wastewater. A typical unit may be 5-8 m in length and 2-3 m in diameter and separated into a series of chambers. The chamber help to maintain a form of plug flow. Aeration occurs as the drum rotates free of the liquid. The large area and good aeration means that the rotating contactor can handle a wide range of flows, needs only short contact times and has 4-5 times the capacity of a conventional filter, and no recycle is required. The disadvantages of the rotating biological contactor are the fact that in cold climates the system needs covering, as well as the cost of running the motor and of maintenance.

Fluidized Bed

One method of increasing the area of the support in a biofilm reactor is to use smaller and robust biofilm supports such as sand. However, sand coated with a biofilm in a reactor would soon clog and trap particles and become anaerobic. To overcome this problem the bed of sand can be fluidized by the upward flow of liquid. The area for biomass support is 3 300 m^2/m^3 compared with 150 m^2/m^3 for rounded gravel which supports an MLSS of 40 000 mg/L compared with 1 500-3 500 mg/L for activated sludge. The high biomass clearly has a considerable oxygen demand which is supplied by injection into the settled sewage of pure oxygen prior to entering the fluidized bed. The biomass build-up on the sand particles can be controlled as coated sand can be extracted, the biomass removed from the sand in a cyclone and the cleaned sand returned to the vessel. This type of system has proved particularly useful for treating high-strength wastes, especially those from industrial sources. The fluidized beds operate with short retention times of around 20 minutes but the system is expensive to operate due to the use of oxygen and the costs of pumping.

An alternative to the use of fluidized beds is the airlift bioreactor where the biomass forms as a biofilm on small particles about 0.3 mm in diameter. The bed is fluidized by the introduction of air at the base of the vessel which also supplies oxygen. The advantage of such a system is that a higher biomass coats the particles due to the better oxygen supply, which means that with the same concentration of organic waste there is a lower growth rate therefore less sludge generated.

Deep Shaft Process

The deep shaft process was a development from ICI's single cell protein work and was based on the airlift design of bioreactor. The airlift design operates by introducing air at the bottom of the vessel. The introduced air will reduce the overall density of the liquid and the air bubbles will rise; these factors combine to cause a flow of water upwards. If this upward flow is separated from the rest of the vessel by a partition (draft tube), a circulating flow will be generated so that both mixing and aeration can be achieved by sparging air. The airlift bioreactor is normally a very tall narrow vessel so that with a height of 100 m or more a pressure of about 10 atmospheres will be found at the base. The high pressure will force more oxygen into solution, improving aeration considerably. In practice the deep shaft bioreactor is sunk in the ground either as concentric pipes or divided vertically and, because of its increased aeration, often has to be installed to treat high BOD industrial wastes. Once a flow has been started, air can also be injected into the downcomer to be carried downwards to the base of the vessel. A deep shaft process has been installed at Marlow Foods' single cell protein plant (Quorn) to treat the waste from the cultivation process, and others have been installed worldwide. The system has the advantage that it requires only a small space compared with conventional systems and due to the high aeration rate will deal with high BOD wastes containing 3-6 kg BOD/m^3·d with a 90 percent treatment rate. This rate is intermediate between the high rate and that of conventional sewage treatment, but the process produces less sludge.

Addition of Pure Oxygen

The aeration of the activated sludge can be improved by the addition of pure oxygen to a closed system or to open tanks. The closed system has the advantage that the oxygen is not lost to the atmosphere, but the presence of a high oxygen concentration does constitute an explosive hazard requiring strict safety precautions. The high aeration also causes an accumulation of carbon dioxide which can reduce the pH and thus reduce nitrification. A system like this was marketed in the UK as the Unox system. The tanks are divided into a series of compartments each of which is mixed by a surface aerator. This type of system can be used to sustain a higher biomass (increased MLSS), with lower sludge production and double the loading rate.

Captor Process

In order to maintain a high biomass at the start of the plug flow process of waste treatment by activated sludge, a modification has been used. Here the activated sludge biomass was immobilized in reticulated plastic pads measuring 25 mm × 25 mm × 12 mm, of a similar nature to washing-up pads. The activated sludge microorganisms which form aggregates readily colonise these pads. The pads are retained in the early part of the aeration tank by screens and have been shown to give higher biomass levels of 6-8 g/L. To maintain an active biomass some of the pads are stripped of excess biomass by a system which removes the pads, squeezes out the sludge and returns the empty pads to the tank.

Membrane Bioreactors

The development of ultrafiltration and microfiltration membranes for biological separations has allowed this technology to be applied to wastewater treatment. The membrane allows the passage of small molecules while retaining the microorganisms making up activated sludge (Brindle and Stephenson, 1996). A membrane bioreactor was first used to treat landfill leachates, but since that time three types of membrane systems have been developed: solid/liquid separation, the gas permeable system and the extractive process.

Membrane bioreactors have been used in both aerobic and anaerobic modes where the high biomass retained gives a rapid breakdown of the organic compounds, enables high loads to be handled and, as the solids are retained, renders the HRT independent of the solids. Because of the high in the system, membrane bioreactors require a good supply of oxygen, but a low substrate to biomass ratio reduces the amounts of the sludge formed. Apart from many experimental system, over 20 full-scale membrane bioreactor units have been installed in the Netherlands and Germany.

As the membrane will allow gases to pass through while retaining the biomass, these bioreactors can provide bubble-free aeration with a high surface area for oxygen transfer. The membrane also provides a support for biofilm formation. In the extractive process the membrane allows chemical pollutants to pass through into the biomass where they can be degraded. The membrane system can also separate the biomass from a biologically hostile waste stream and this type of system has been tested with pollutants such as nitrobenzene, benzene and dichloroaniline with over 99 percent removal. Membrane bioreactors are more expensive than conventional activated sludge and trickling filter processes but have the advantage that less sludge is generated, they have a high COD removal and good oxygen supply and appear to be suitable for small plants and where high-quality effluent is required.

Vocabulary

biotower	生物塔	immobilize	固定
rotating biological contactor	生物(接触)转盘	landfill leachate	垃圾渗滤液
biofilm reactor	生物膜反应器	nitrobenzene	硝基苯
high-strength	高浓度的	benzene	苯
deep shaft process	深井工艺	dichloroaniline	二氯苯胺
fluidized bed	流化床		

Reading Material B

Sludge Treatment

Conventional activated sludge and filter processes produce large volumes of primary sludge in addition to the excess settled secondary sludge (activated sludge). Almost 1 million tones (dry weight) are produced each year in the UK. In the activated sludge process this secondary sludge is mainly the microbial biomass produced by the metabolism of the organic material. The microbial yield on settled sewage is about 50 percent. A proportion of this biomass is recycled (20 percent) and the remainder is combined with the primary sludge for disposal.

In the trickling filters the same problem applies except that with a lower loading less sludge is produced but there is no recycle. Therefore, large volumes of sludge are formed which have a solids content of about 1-4 percent and represent one of the main problems of disposal in wastewater treatment. The waste sludge is a mixture of organic material and microbial cells which can be degraded by other microorganisms. There are a number of methods employed to dispose of excess sludge. These are:

- dumping at sea;
- landfill;
- incineration;
- spray irrigation (agricultural disposal);
- drying;
- composting;
- anaerobic digestion.

In the UK until recently 67 percent of the sludge was disposed of on land, 29 percent dumped at sea, and 4 percent removed by incineration. It was agreed at the 1990 North Sea Conference to phase out disposal of sewage sludge at sea by 1998 to comply with EC Urban Wastewater Treatment Directive 91/271/EEC. The banning of sludge dumping at sea was a measure to improve coastal water quality, although there was no substantial evidence of environmental damage other than chromium accumulation. Thus there will be a need to find a replacement for dumping at sea.

Of the 67 percent sludge disposed on land some 16 percent goes into landfill sites, either solely or co-disposed with domestic wastes in the UK. A much higher proportion is landfilled in the UK than in other European countries where incineration and composting are used more widely. In the USA in 1993 62 percent of domestic waste was landfilled, 16 percent combusted and 4 percent composted.

Despite recent reductions in landfill use in the USA it is clear that most of the waste is placed in landfill. However, of the 6 000 landfill sites used in the USA in 1991, 3 000 will be closed by the year 2000. This will mean that new sites will be needed but there is considerable public resistance to the formation of new sites.

The large-scale incineration of sewage sludge is expensive, with high capital costs, and is only a partial disposal option as the ash formed needs disposal. However, the development of autothermic incineration has made incineration more attractive. In the process primary and secondary sludges are mixed together and water removed by pressing so that a cake of 30 percent solids is produced which can support autothermic combustion. Advanced combustion systems such as fluidized beds working at temperatures of 750-850 ℃ create more heat than is required to heat the inlet air and remove the water from the sludge. This means that once the process has started, no fuel needs to be added as the sludge itself generates sufficient heat. The ash formed can be removed by an electrostatic precipitator and a wet scrubber will remove sulphur dioxide, hydrogen fluoride and hydrogen chloride. The ash contains the heavy metals which are present in the sludge and represents 30 percent of the original dry mass and 1-2 percent of the volume and is normally disposed of in landfill sites.

Some sludge is disposed of by some form of agricultural use (Wheatly, 1985). Any sludge which is applied to agricultural land is required to have some form of biological or chemical treatment to reduce the levels of pathogens unless it is injected below the surface. These treatments are:

- pasteurization, heating at 70 ℃ for 30 min or more;

- anaerobic digestion at 35 ℃ for 12 days;
- thermophilic aerobic digestion, 7 days at 55 ℃;
- composting;
- alkali stabilisation, pH > 12 for 2 hours;
- liquid storge, 3 months;
- drying and storage, 3 months.

All these methods reduce the pathogen content of the sludge before it is applied to the land, but there are restrictions on what crops are grown and how soon the land can be used for certain crops. Treated sludge can be applied to land to be used for growing fruit vegetable crops until 10 months after application. Prior to drying, the sludge is normally conditioned so that its ability to settle before dewatering is improved. Often polyvalent ions such as Fe^{2+} or Al^{3+}, polyelectrolytes or soil is added to improve precipitation. Dewatering is often carried out in drying beds but filtration and centrifugation have also been used to produce a compact cake.

Almost all sludge contains heavy metals, as microorganisms have the ability to sequester metals, so that the application of sludge to soils carries with it the risk of producing high levels of heavy metals in the soil. There are limits for metals in sewage sludge and in soil.

Vocabulary

metabolism	新陈代谢	combustion	燃烧
dumping at sea	填海	thermophilic	喜温的,嗜热的,适温的
incineration	焚烧	autothermic	热自动补偿的
composting	堆肥	pasteurization	加热杀菌法,巴斯德杀菌法
sulphur	硫磺	polyvalent	聚合价的

Unit 23 Microbes as Chemical Machine

In 1857 Louis Pasteur published a report showing that the souring of milk is caused by microbes which convert milk sugar into lactic acid. The process can be chemically expressed as follows:

$$\underset{\text{(sugar)}}{C_6H_{12}O_6} \longrightarrow \underset{\text{(lactic acid)}}{2CH_3CHOHCOOH}$$

In its gross result and in appearance at least, this is one of the simplest chemical changes that can be imagined, representing as it does the breakdown of one molecule of sugar into two molecules of lactic acid.

Let us now turn to another transformation of sugar studied by Pasteur during the same period, namely alcoholic fermentation as exemplified by the conversion of grape juice into wine. In this case, as is well known, the microorganism responsible for the change is yeast, which converts the sugar of grape juice into alcohol according to the following equation:

$$\underset{\text{(sugar)}}{C_6H_{12}O_6} \longrightarrow \underset{\text{(alcohol)}}{2CH_3CH_2OH} + 2CO_2$$

The Presence or Absence of Oxygen

When left exposed to the air, wine will turn into vinegar, and this is precisely the next problem to which Pasteur addressed himself. He showed that in this case the change is caused by still another type of microorganism, which oxidizes alcohol into acetic acid according to the following formula:

$$\underset{\text{(alcohol)}}{2CH_3CH_2OH} + O_2 \longrightarrow \underset{\text{(acetic acid)}}{CH_3COOH} + H_2O$$

As we have just seen, the conversion of sugar into lactic acid or alcohol occurs independently of the presence of oxygen, whereas the conversion of alcohol into acetic acid results from an oxidation in which atmospheric oxygen participates. In contrast, Pasteur observed that when a sugar solution was placed in an atmosphere from which the oxygen had been completely removed a very different kind of substance was likely to appear, namely butyric acid. Under these conditions, the bacteria which proliferate live best without oxygen and in fact may die in the presence of this gas. "Anaerobic" bacteria (to use an expression suggested by Pasteur himself) convert sugar into butyric acid according to the following general formula, which corresponds to "anaerobic" fermentation, that is, one not involving the use of oxygen.

$$\underset{\text{(sugar)}}{C_6H_{12}O_6} \longrightarrow \underset{\text{(butyric acid)}}{CH_3CH_2CH_2COOH}$$

The World of Microbes

I have listed four different types of chemical processes, not only to demonstrate therange of Pasteur's contributions to microbial chemistry, but even more to illustrate the chemical versatility of microorganisms. The important lesson to be learned from these simple examples is that given a certain substance, it can be transformed into many different derivative substances depending upon the types of microbes to which it is exposed, and upon the particular conditions under which microbial action takes place.

Pasteur discovered that each type of microbe is more or less specialized in a few chemical reactions; and for every kind of organic substance there exists in nature at least one and usually several kinds of microbes capable of attacking and breaking down this substance provided the conditions are right. The many types of microorganisms that are present almost everywhere in nature break down all complex organic substances step by step into simpler and simpler chemical compounds. They are necessary links in the endless chain which binds life to matter and matter to life. Without them life would soon come to an end.

Further evidence of the immense part that microorganisms play in the world of nature is found in the fact that, according to recent calculations, the total mass of microbial life on earth is approximately twenty times greater than the total mass of animal life!

During recent decades scientists working on microbial nutrition have found that vitamins provide still further evidence of similarity among the different forms of life from the largest to the smallest. It must be said in truth that many microorganisms can grow in culture media lacking sufficient vitamin for the survival and growth of animals and men, but the reason is that these microorganisms can themselves synthesize the vitamins from simpler materials and thus produce enough to satisfy their own needs. In other words, whether microorganisms do or do not require preformed vitamins for growth, they utilize these substances for their biochemical activities in much the same way as do plants, animals, and men.

Vocabulary

lactic acid	乳酸	butyric	丁酸的
breakdown	分解	proliferate	繁殖,繁衍
fermentation	发酵	versatility	多面性,多功能性
equation	反应式,公式	media	培养基
acetic	乙酸的,醋酸的	biochemical	生物化学

Reading Material A

Introduction of Sewage Treatment

The use of microorganisms in the disposal of sewage was developed around 1910 ~ 1914 in Manchester and a number of continental European cities. In earlier days, sewage was buried or run into rivers and waterways. In the nineteenth century the populations of most cities expanded greatly due to industrialization, so that the volume of sewage produced was too much for this type of disposal and rivers and canals became very polluted. In the UK rivers such as the Thames became anaerobic and devoid of aquatic life, producing unpleasant smells, and they were largely responsible for the spread of diseases such as typhoid and cholera. Sewage pollution was so acute in London that it was the practice for those who could afford it to move to the country in the summer. The treatment of sewage is linked to the provision of clean water, as polluted water is cause of many disease. The elimination or reduction of waterborne diseases by the introduction of sewers and sewage disposal probably did more for the health of the population than the introduction of antibiotics later in the century.

Pollution

The disposal of sewage by running it into the sea or a waterway allows the indigenous microbial population to degrade the waste. The dilution of the waste meant that the organic content of the waterway did mot reach too high a value. Natural waterways contain a population of microorganisms that utilize dissolved organic compounds, that in turn are part of the food chain for protozoan, insects, worms and fish. Under normal conditions in a waterway, this population of organisms forms a balanced ecosystem. The balanced system can be destabilized by the addition of excess metabolic organic material, such as high levels of sewage. The addition of metabolic organic compounds causes a considerable increase in the growth and metabolism of the aerobic microbial population of the waterway which will use all available oxygen dissolved in the water.

The dissolved oxygen level drops rapidly at the point of addition (often known as BOD sag), but provided the amount of organic material added is not too high, the oxygen level will rise slowly as the material moves down-stream from the addition site. The subsequent rise in dissolved oxygen is due in part to mixing and dilution with unaffected flowing water in the waterway and the metabolism of the organic material. However, if the organic addition is too great, the increased microbial metabolism keeps the conditions in the waterway anaerobic. The anaerobic condition, if it persists, causes the aerobic microorganisms to decline or die and allows the anaerobes to increase, anaerobic metabolism is slower than aerobic metabolism, so that the rate of degradation of the organic material decreases, which can lead to build-up of excess organic material. The continued lack of oxygen leads to the death of other aquatic organisms such as fish and crustacean. In addition, the anaerobic metabolism produces gases such as hydrogen sulphide and methane which can be an indication of the anaerobic condition in the waterway. Stagnant waters such as lacks and ponds become anaerobic more rapidly on the addition of excess organic material as they have no flow to mix the system and to add clean water.

Domestic waste comes in two forms—liquid or solid. Domestic wastewater is derived from private homes, commercial buildings and institutions such as schools and hospitals. In some cases it may have additions of industrial wastewater from processes such as brewing, baking, paper mills and metal processing. In general, however, treatment plants normally accept only domestic wastewater, and industrial plants have to treat their own waste prior to returning it to waterways or to the domestic wastewater system. If an industrial waste is discharged to the sewers, the water company normally makes a charge for its treatment.

Waterways are affected not only by the addition of biodegradable wastes (nutritional pollution) but also frequently by chemical and physical pollution. Chemical pollution is mainly industrial with the release of acids, alkalis and toxic compounds which can poison the living organisms in the waterway. Physical pollution is the release into the waterway of materials which can change the physical condition in the waterway. This type of pollution mainly consists of the release of large quantities of warm water from power stations which use large volumes of water for cooling. The change in temperature of the water can encourage excess growth of native organisms or growth of a new organism, thus changing the balance of the normal population.

Sewage

Sewage consists of some 99.99 percent by weight of water containing dissolved organic material, suspended solids, microorganisms (pathogens) and a number of other components. The suspended solids can rang from > 100 μm to colloidal in nature. The composition, concentration and condition of the waste may differ widely depending on the origin, time or weather condition. The strength can vary daily and with season and can be diluted by rainwater.

In typical sewage, 75 percent of the suspended solids and 40 percent of the dissolved material are organic. The dissolved organic materials are a mixture of proteins, carbonhydrates, fat and detergents. There are substantial inorganic components in sewage, including sodium, calcium, magnesium, chlorine, sulphates, phosphates, bicarbonates, nitrates and ammonia with races of heavy metal. The oxygen demand created by the organic material in sewage is usually expresses as BOD (Biological Oxygen Demand) in g/m^3 or mg/L. The oxygen demand of the sewage can also be estimated by means of a chemical technique using an oxidizing agent and this is known as the chemical oxygen demand (COD). Sewage normally has a BOD value of 200-600 mg/L, whereas wastes from some industrial and agricultural processes such as tannery waste and animal slurry can have much higher BOD values of up to 50,000 mg/L.

Function of Waste Treatment Systems

The main function of domestic waste treatment systems is to reduce the organic content as far as possible in order to be able to return the water to rivers and coastal waters without causing nutritional pollution, especially to those rivers used as a source of the drinking water. In addition, the system should remove suspended matter, reduce pathogen content, and increasingly remove nitrates, heavy metal and man made chemicals.

The quality of the treated waste released from the treatment system depends on the volume and condition of the receiving water and its ability to dilute the waste, and on whether there is to be water abstraction further downstream. In general a 20:30 standard is adopted, which is an effluent level of 20 mgBOD_5/L and 30 mg/L suspended solids(SS) when the effluent is to be diluted 8:1.

The volume of domestic wastewater from toilets, baths, washing machines etc, in the UK is about 9 million m^3/day, industry uses about 7 million m^3/day and with average run-off due to rainfall this gives a total of about 18 million m^3/day, all of which needs treating before it can return to the rivers or canals. It has been calculated that each person in the UK contributes on average some 230 litres of sewage per day. Thus, although the scale of sewage treatment is enormous, it is a very dilute growth medium and its value per tone is very low compared with other biotechnological processes (Wheatly, 1985).

Vocabulary

disposal	处理,处置
industrialization	工业化,产业化
aquatic	水的,水上的,水生的,水栖的
protozoan	原生动物
metabolism	新陈代谢
hydrogen sulphide	硫化氢
stagnant	停滞的,迟钝的
bicarbonate	重碳酸盐
abstraction	提取
dilute	稀释,冲淡

Reading Material B

Biological Treatment System

Biological processing is the most efficient way of removing organic matter from municipal waste waters. These living systems rely on mixed microbial cultures to decompose, and to remove colloidal and dissolved organic substances from solution. The treatment chamber holding the microorganisms provides a controlled environment; for example, activated sludge is supplied with sufficient oxygen to maintain an aerobic condition. Waste water contains the biological food, growth nutrients, and inoculum of microorganisms. Persons who are not familiar with waste water operations often ask where the "special" biological cultures are obtained. The answer is that the wide variety of bacteria and protozoa present in domestic wastes seed the treatment units. Then by careful control of waste water flows, recirculation of settled microorganisms, oxygen supply, and other factors, the desirable biological cultures are generated and retained to process the pollutants. The slime layer on the surface of the media in a trickling filter is developed by spreading waste water over the bed. Within a few weeks the filter is operational, removing organic matter from the liquid trickling through the bed.

Activated sludge in a mechanical, or diffused-air, system is started by turning on the aerators and feeding the waste water. Initially a high rate of recirculation from the bottom of the final clarifier is necessary to retain sufficient biological culture. However, within a short period of time a settleable biological floc matures that efficiently flocculates the waste organics. An anaerobic digester is the most difficult treatment unit to start up, since the methane forming bacteria, essential to digestion, are not abundant in raw waste water. Furthermore, there anaerobic grow very slowly and require optimum environmental conditions. Start-up of an anaerobic digester can be hastened considerably by filling the tank with waste water and seeding with a substantial quantity of digesting sludge from a nearby treatment plant. Raw sludge is then fed at a reduced initial rate, and lime is supplied as necessary to hold pH. Even under these conditions, several months may be required to get the process fully operational.

The most important factors affecting biological growth are temperature, availability of nutrients, oxygen supply, pH, presence of toxins and, in the case of photosynthetic plants, sunlight. Bacteria are classified according to their optimum temperature range for growth. Mesophilic bacteria grow in a temperature range of 10 to 40 ℃, with an optimum of 37 ℃. Aeration tanks and trickling filters generally operate in the lower half of this range with wastewater temperatures of 20 to 25 ℃ in warm climates and 8 to 10 ℃ during the winter in northern regions.

Municipal waste waters commonly contain sufficient concentrations of carbon, nitrogen, phosphorus, and trace nutrients to support the growth of a microbial culture. Theoretically, a BOD to nitrogen to phosphorus ratio of 100∶5∶1 is adequate for aerobic treatment, with small variations depending on the type of system and mode of operation. Average domestic waste water exhibits a surplus of nitrogen and phosphorus with a BOD∶N∶P ratio of about 100∶17∶5. If a municipal waste contains a large volume of nutrient-deficient industrial waste, supplemental nitrogen is generally supplied by the addition of anhydrous ammonia (NH_3) or phosphoric acid (H_3PO_4) as is needed.

Diffused and mechanical aeration basins must supply sufficient air to maintain dissolved oxygen for the biota to use in metabolizing the waste organics. Rate of microbial activity is independent of dissolved oxygen concentration above a minimum critical value, below which the rate is reduce by the limitation of oxygen required for respiration. The exact minimum depends on the type of activated sludge process and the characteristics of the waste water being treated. The most common design criterion for critical dissolved oxygen is 2.0 mg/L, but in actual operation values as low as 0.5 mg/L have proved satisfactory. Anaerobic systems must, of course, operate in the complete absence of dissolved oxygen; consequently, digesters are sealed with floating or fixed covers to exclude air.

Hydrogen ion concentration has a direct influence on biological treatment systems which operate best in a neutral environment. The general range of operation of serration systems is between pH 6.5 and 7.4. Fungi are favored over bacteria in the competition for metabolizing the waste organics. Anaerobic digestion has a small pH tolerance range of 6.7 to 7.4 with optimum operation at pH 7.0 to 7.1. Domestic waste sludge permits operation in this narrow range except during start-up or periods of organic overloads. Limited success in digester pH control has been achieved by careful addition of lime with the raw sludge feed. Unfortunately, the buildup of acidity and reduction of pH may be a symptom of other digestion problems, for example, accumulation of toxic heavy metals which the addition of lime cannot cure.

Biological treatment systems are inhibited by toxic substances. Industrial wastes from metal finishing industries often contain toxic ions, such as nickel and chromium; chemical manufacturing produces a wide variety of organic compounds that can adversely affect microorganisms. Since little can be done to remove or neutralize toxic compounds in municipal treatment, pretreatment should be provided by industries prior to discharging wastes to the city sewer.

Vocabulary

inoculum	接种体	neutral	中性的
decompose	分解	symptom	症状，征兆
start-up	启动	adversely	相反地，反对地
mesophilic	(细菌)嗜温的		

Unit 24 Nitrification and Denitrification

Waste streams contain not only biologically metabolisable organic materials but also nitrogen and phosphorus-containing compounds. The nitrogen-containing compounds in sewage are ammonia, proteins and amino acids. Increasingly nitrates are being found in wastewater as a result of run-off from agricultural land. Ammonia is formed from urea, a major constituent of urine, and can also be formed during the natural breakdown of proteins and can be found in some industrial wastes. Ammonia has an offensive smell, is poisonous to aquatic life at concentrations as low as 0.5 mg/L, and increases the chlorine dosage needed for the treatment of drinking water. The European Inland Fisheries Advisory Commission (EIFAC) has recommended that the maximum ammonia concentration should be 0.025 kg/m^3 (20 g/m^3 total ammonia).

Nitrogen-containing compound are required for the synthesis of proteins by all types of microorganisms during their growth. The microorganisms normally utilize ammonia first but nitrates and urea can also be used. In waste treatment systems a third of the total nitrogen content is removed by assimilation during growth, and the rate of removal depends on the biomass level and rate of growth in the waste treatment system.

Nitrification

The level of ammonia in wastewater (domestic) is about 25 mg/L (g/m^3) but industrial wastes can contain up to 5 g/L ammonia with the recommended level at 20 mg/L (25 g/m^3). Ammonia is oxidized rapidly to nitrate in the environment and in wastewater treatment systems in a process known as nitrification. The conversion is carried out by two groups of chemoautotrophic bacteria which use the oxidation of ammonia as a source of energy, the first stage of ammonia oxidation is carried out mainly by the genera *Nitrosomonas* and *Nitrosococcus*, *Nitrosospira*, and *Nitrocystis*. The reaction is as follows, although the oxidation of ammonia is more complex than is given in the equation:

$$2NO_4^+ + 3O_2 \longrightarrow 2NO_2 + 4H^+ + 2H_2O + (\text{energy } 480\text{-}700 \text{ kJ})$$

The energy released is used by the organisms to synthesise cell components from inorganic sources. The release of hydrogen ions can cause a drop in pH and it is clear that a good supply of oxygen is required. The growth of nitrifying bacteria is very slow (u_{max}0.46-2.2/d) compared with that of heterotrophic bacteria (u_{max}0.1-1/d).

The nitrite formed is converted to nitrate by the genera *Nitrobacter*, *Nitrocystis*, *Nitrosococcus* and *Nitrosocystis*, but *Nitrobacter* has been the most studied. The reactions are as follows:

$$2NO_2^- + O_2 \longrightarrow 2NO_3^- + (\text{energy } 130\text{-}180 \text{ kJ})$$

As the oxidation of nitrite to nitrate yields less energy than the oxidation of ammonia, the cell yield of *Nitrobacter* is less than that of *Nitrosomonas* and the growth rates are also slow with a u_{max} of 0.28-1.44/d. The characteristics of the organisms involved in nitrification affect the wastewater treatment as follows:

- The growth rate is slower than that of heterotrophic organisms so that the organic load has to be balanced to their slower growth rate, otherwise the organisms will be washed out.
- There is a low cell yield per unit of ammonium oxidised.
- The organisms require a significant amount of oxygen, 4.2 g per g NH_4 converted.
- The system may need some form of buffering due to the acid conditions produced by the hydrogen ions.

Denitrification

Nitrification in treatment plants and soils combined with the nitrates from agricultural run-off can give rise to high nitrate levels (above 50 mg/L) in waterways which are used to supply drinking water. High nitrate levels are associated with one disease, which affects children below the age of 6 months. The children have an incomplete digestive system and the intake of nitrate leads to the accumulation of nitrite ions which enter the blood system and block haemoglobin oxygen transport. Thus the EC have set an absolute limit of 50 mg/L for nitrate, and a recommended limit of 25 mg/L, although a number of UK water treatment systems are working at 80 mg/L. Nitrate can be converted to nitrite in the human stomach and nitrite has been shown to be converted to carcinogenic nitrosamines, leading to concern over the development of stomach cancer on consumption of high-nitrate water.

Nitrate removal or denitrification can be carried out by ion exchange or biological processes. The ion exchange process depends on the resin's affinity which on a conventional anion exchange resin is:

$$SO_4^{2-} \gg NO_3^- > Cl^- \geqslant HCO_3^-$$

Any sulphate in the waste will bind in preference to nitrate, but once this has occurred nitrate will exchange with chloride. Once the resin is exhausted it will require regeneration with excess sodium chloride which yields a solution containing high concentrations of sodium sulphate, sodium nitrate and sodium chloride, which will need disposal. Adding this high-salt solution to waterways is unacceptable and in practice it is passed onto the sewage works for treatment.

The biological conversion of nitrate to nitrite and eventually nitrogen occurs under conditions where oxygen is very low or absent. The process of oxidation involves the loss of electrons and in normal conditions oxygen acts as an electron acceptor, but when oxygen levels are low inorganic ions such as nitrate, phosphate and sulphate can act as electron acceptors. In wastewater where nitrification has occurred, combined with nitrate from agricultural run-off, the concentration of nitrate will be higher than that of sulphate or phosphate. A number of facultative heterotrophic microorganisms occur in sewage treatment system which are capable of concerting nitrate to nitrogen provided an electron donor and the electron donor is present. The electron donor is usually an organic compound and in some cases methanol has been used to supplement the normal organic source. The reactions with methanol are as follows:

$$3NO_3^- + CH_3OH \longrightarrow 3NO_2^- + CO_2 + 2H_2O$$

$$2NO_2^- + CH_3OH \longrightarrow N_2 + CO_2 + H_2O + 2OH^-$$

The process of denitrification requires low oxygen levels (anaerobic), an organic carbon energy source, a level of nitrate of 2 mg/L or above, and a pH of 6.5 to 7.5.

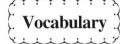

Vocabulary

protein	蛋白质
amino acid	氨基酸
chemoautotrophic	化能自养的,化学自养的
Nitrosomonas	亚硝化胞菌属
Nitrosococcus	亚硝化球菌属
Nitrosospira	亚硝化螺菌属
Nitrocystis	硝化囊菌属
heterotrophic	异养的
haemoglobin	血色素,血红蛋白
carcinogenic	致癌物的
nitrosamine	亚硝氨

Reading Material A

Nitrification and Denitrification Processes

Within the sewage system the processes of biological nitrification and denitrification can be organized in a number of ways. In general, the first step, the removal of ammonia by nitrification, can be carried out in parallel with the removal of organic material, provided that the hydraulic retention time is not too short (Tab. 24.1). Denitrification, in contrast, requires a change in growth conditions from aerobic to anaerobic and an organic carbon source. Both nitrification and denitrification can be achieved by partitioning the sewage treatment system (Fig. 24.1) (Lee et al., 1997) or by providing separate reactors which can be used with both suspended and fixed film cultures (Fig. 24.2) (Tchobanoglous and Burton, 1991). In the single vessel the anoxic zone is situated at the start of the aeration tank where anoxic conditions are achieved by stopping aeration and the carbon level is high. The process using separate vessels is much easier to use and control.

Tab. 24.1 Parameters for biological nitrification/denitrification process

Process	Sludge retention time /d	retention time T /h	MLVSS /($g \cdot L^{-1}$)	pH
Carbon removal	2 – 5	1 – 3	1 – 2	6.5 – 8.0
Nitrification	10 – 20	0.5 – 3	1 – 2	7.4 – 8.6
Denitrification	1 – 5	0.2 – 2	1 – 2	6.5 – 7.5

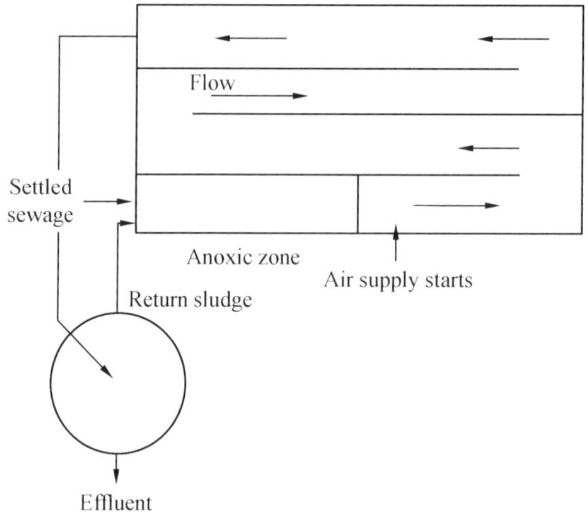

Fig. 24.1 An activated sludge process with the provision of an anoxic zone at the start in order to achieve denitrification

Note: The anoxic zone is formed by stopping aeration at the first stage where the organic material is highest.

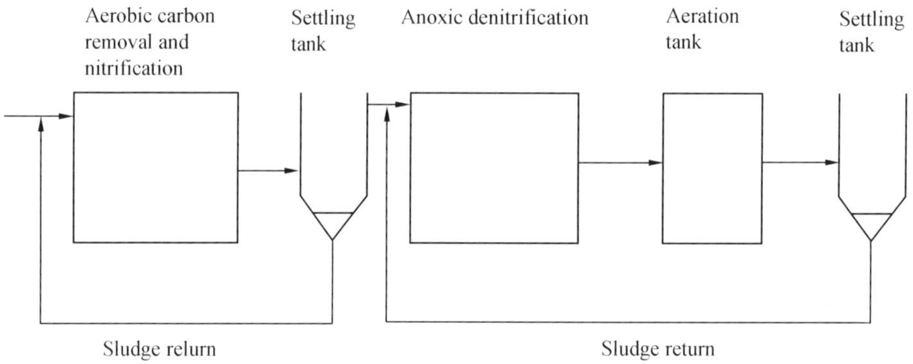

Fig. 24.2 An outline of a two stage process for both nitrification and denitrification

Note: In the first stage the normal activated sludge process occurs with the removal of organic materials and nitrification. In the second stage anoxic conditions cause denitrification which is followed by an aeration tank to strip out the nitrogen formed to ensure precipitation in the settling tank.

Another system for combined nitrification and denitrification is the sequencing batch reactor (SBR) where a single vessel is used but a programmed sequence of operations are applied which can be feeding, anaerobic conditions, aerobic conditions, sludge settling, and effluent removal. This type of operation has been used for the treatment of a number of wastes such as agricultural run-off and landfill leachates, but it has the potential for combining nitrification and denitrification. An example of nitrate, nitrite, and ammonia levels in a sequencing batch reactor is given in Fig. 24.3. The ammonia levels drop during the initial anaerobic phase and the subsequent aerobic phase. Nitrate concentration in contrast is low at the start but rises due to nitrification in the aerobic phase. Both nitrate and nitrite were denitrified during the anoxic phase. The sequencing mode can also be applied to anaerobic reactors (Ndon and Dague, 1997) and in two-stage processes (Ra et al., 1998) where the second stage is the anoxic phase.

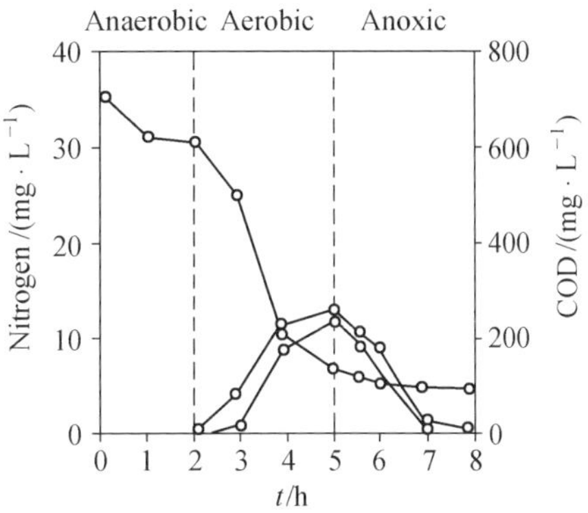

Fig. 24.3 The levels of ammonia(○), nitrate(□), and nitrite (△) in a sequencing batch reactor

Note: The three sequence of anaerobic, aerobic and anoxic are followed by a 1.5 hour settlement phase and a 0.5 hour decantation.

Vocabulary

hydraulic retention time	水力停留时间	nitrate	硝酸盐
organic carbon source	有机碳源	nitrite	亚硝酸盐
partition	分开,分隔		

Reading Material B

Biological Phosphorus Removal

The removal of phosphorus by biological means is known as biological phosphorus removal. Phosphorus removal is generally done to control eutrophication because phosphorus is a limiting nutrient in most freshwater systems. Treatment plant effluent discharge limits have ranged from 0.10 to 2.0 mg/L of phosphorus depending on plant location and potential impact on receiving waters. Chemical treatment using alum or iron salts is the most commonly used technology for phosphorus removal, but since the early 1980s success in full-scale plant biological phosphorus removal has encouraged further use of the technology. The principal advantages of biological phosphorus removal are reduced chemical costs and less sludge production as compared to chemical precipitation.

Process Description

In the biological removal of phosphorus, the phosphorus in the influent wastewater is incorporated into cell biomass, which subsequently is removed from the process as a result of sludge wasting. Phosphorus accumulating organisms (PAOs) are encouraged to grow and consume phosphorus in systems that use a reactor configuration that provides PAOs with a competitive advantage over other bacteria. The reactor configuration utilized for phosphorus removal is comprised of an anaerobic tank having a value of 0.50 to 1.0 h that is placed ahead of the activated sludge aeration tank (see Fig. 24.4). The contents of the anaerobic tank are mixed to provide contact with the return activated sludge and influent wastewater. Anaerobic contact tanks have been placed in front of many different types of suspended growth processes, with aerobic SRT value ranging from 2 to 40 d.

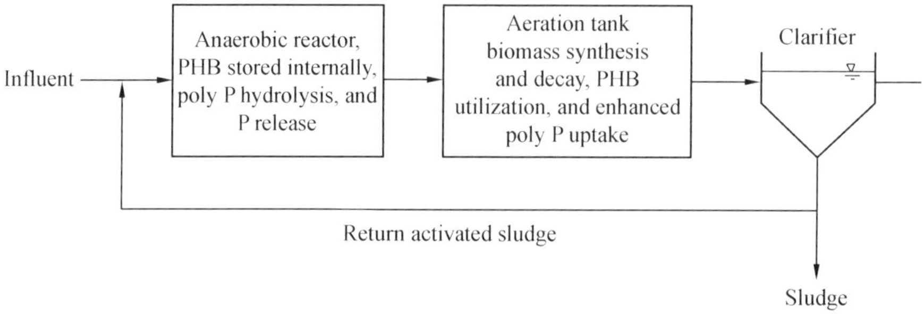

Fig. 24.4 Biological phosphorus removal: typical reactor configuration

Phosphorus removal in biological systems is based on the following observations (Sedlak, 1991):

1. Numerous bacteria are capable of storing excess amounts of phosphorus as poly-phosphates in their cells.

2. Under anaerobic conditions, PAOs will assimilate fermentation products (e.g., volatile fatty acids) into storage products within the cells with the concomitant release of phosphorus from stored polyphosphates.

3. Under aerobic conditions, energy is produced by the oxidation of storage products and polyphosphate storage within the cell increases.

A simplified version of the process occurring in the anaerobic and aerobic/anoxic reactors or zones is presented below. In many applications for phosphorus removal, an anoxic reactor follows the anaerobic reactor and precedes the aerobic reactor. Most PAOs can use nitrite in place of oxygen to oxidize their stored carbon source. A more comprehensive description of the biochemistry and intracellular transformations can be found in Wentzel et al. (1991).

Processes Occurring in the Anaerobic Zone

- Acetate is produced by fermentation of COD which, as defined earlier, is dissolved degradable organic material that can be assimilated easily by the biomass. Depending on the value of t for the anaerobic zone, some colloidal and particulate COD is also hydrolyzed and converted to acetate, but the amount is generally small compared to that from the COD conversion.

- Using energy available form stored polyphosphates, the PAOs assimilate acetate and produce intracellular polyhydroxybutyrate (PHB) storage products. Some glycogen contained in the cell is also used. Concurrent with the acetate uptake is the release of orthophosphate (O-PO_4), as well as magnesium, potassium, and calcium cations.

● The PHB content in the PAOs increases while the polyphosphate decreases.

Processes Occurring in the Aerobic/Anoxic Zone

● Stored PHB is metabolized, providing energy from oxidation and from carbon for new cell growth.

● Some glycogen is produced from PHB metabolism.

● The energy released from PHB oxidation is used to form polyphosphate bonds in cell storage so that soluble orthophosphate (O-PO_4) is removed from solution and incorporated into polyphosphates with the bacterial cell. Cell growth also occurs due to PHB utilization and the new biomass with high polyphosphate storage accounts for phosphorus removal.

● As a portion of the biomass is wasted, stored phosphorus is removed from the biotreatment reactor for ultimate disposal with the waste sludge.

Microbiology

Phosphorus is important in cellular energy transfer mechanisms via adenosine triphosphate (ATP) and polyphosphates. As energy is produced in oxidation reduction reactions, adenosine diphosphate (ADP) is converted to ATP with 1.69 kJ/mol energy captured in the phosphate bond. As the cell uses energy, ATP is converted to ADP with phosphorus release. For common heterotrophic bacteria in activated-sludge treatment the typical phosphorus in their cells in the form of energy-rich polyphosphates, resulting in phosphorus content as high as 20 to 30 percent by dry weight.

In the anaerobic zone, concentrations of O-PO_4 as high as 40 mg/L can be measured in the liquid, as compared to wastewater influent concentrations of 5 to 8 mg/L. The high concentration of O-PO_4 can be taken as an indication that phosphorus release by the bacteria has occurred in this zone. Also in this zone, significant amounts of poly-hydroxybutyrate (PHB) are found stored in bacteria cells, but the PHB concentration declines appreciably in the subsequent anoxic and /or aerobic zones and can be measured and quantified. The O-PO_4 is taken up from solution in the aerobic and anoxic zones, generally leading to very low remaining concentrations. Based on investigations of biological phosphorus removal, it was found that acetate was essential to forming the PHB under anaerobic conditions, which provide a competitive advantage for the PAOs.

The anaerobic zone in the anaerobic/aerobic treatment process is termed a "selector". Because it provides conditions that favor the proliferation of the PAOs, by the fact that a portion of the influent bCOD is consumed by the PAOs instead of other heterotrophic bacteria. Because the PAOs prefer low-molecular-weight fermentation product substrates, the preferred food source would not be available without the anaerobic zone that provides for the fermentation of the influent bsCOD to acetate. Because of the polyphosphate storage ability, the PAOs has energy available to assimilate the acetate in the anaerobic zone. Other aerobic heterotrophic bacteria have no such mechanism for acetate uptake, and they are starved while the PAOs assimilate COD in the anaerobic zone. It should also be noted that the

PAOs form very dense, good settling floc in the activated sludge, which is an added benefit. In some facilities, the anaerobic/aerobic process sequence has been used because of the sludge settling benefits, even though biological phosphorus removal was not required.

Care must be taken in the handling of the waste sludge from biological phosphorus removal systems. When the sludge is held under anaerobic conditions, phosphorus release will occure. Release of O-PO_4 is possible even without acetate addition as the bacteria use the stored polyphosphate for an energy source. The release of O-PO_4 can also occur after extended contact time in the anaerobic zone of the biological phosphorus treatment system. In that case the released phosphorus may not be taken up in the aerobic zone because the release was not associated with acetate uptakes and PHB storage for later oxidation. The release of O-PO_4 under these conditions is termed secondary release (Barnard, 1984), which can lead to a lower phosphorus removal efficiency for the biological process.

Vocabulary

biological phosphorus removal	生物除磷
phosphorus accumulating organism	聚磷生物
SRT	污泥停留时间
polyphosphate	聚磷
volatile fatty acid	挥发脂肪酸
triphosphate	三磷酸
diphosphate	二磷酸
poly-hydroxybutyrate	聚烃基丁酸
orthophosphate	正磷酸盐

Unit 25 Advanced Wastewater Treatment

Advanced wastewater treatment is defined as the additional treatment needed to remove suspended, colloidal, and dissolved constituents remaining after conventional secondary treatment. Dissolved constituents may range from relatively simple inorganic ions, such as calcium, potassium, sulfate, nitrate, and phosphate, to an ever-increasing number of highly complex synthetic organic compounds. In recent years, the effects of many of these substances on the environment have become understood more clearly. Research is ongoing to determine 1) the environmental effects of potential toxic and biologically active substances found in wastewater and 2) how these substances can be removed by both conventional and advanced wastewater treatment processes. As a result, wastewater treatment are becoming more stringent in terms of both limiting concentrations of many of these substances in the treatment plant effluent and establishing whole effluent toxicity limits. To meet these new requirements, many of the existing secondary treatment facilities will have to be retrofitted and new advanced wastewater treatment facilities will have to be constructed. Therefore, the purpose of this unit is to provide an introduction to the subject of advanced wastewater treatment. The unit contains an expanded discussion of the need for advanced wastewater treatment, an overview of the available technologies used to the removal of the constituents of concern, and an introduction to the more important of these technologies as applied to the removal of specific constituents found in wastewater.

Need for Advanced Wastewater Treatment

The need for advanced wastewater treatment is based on a consideration of one or more of the following factors.

1. The need to remove organic matter and total suspended solids beyond what can be accomplished by conventional secondary treatment processes to meet more stringent discharge and reuse requirements.

2. The need to remove residual total suspended solids to condition the treated wastewater for more effective disinfection.

3. The need to remove nutrients beyond what can be accomplished by conventional secondary treatment processes to limit eutrophication of sensitive water bodies.

4. The need to remove specific inorganic and organic constituents to meet more stringent discharge and reuse requirements for both surface water and land-based effluent dispersal and for indirect potable reuse applications.

5. The need to remove specific inorganic (e.g., heavy metals, silica) and organic constituents for industrial reuse.

With increased scientific knowledge derived from laboratory studies and environmental monitoring concerning the impact of the residual constituents found in secondary effluent, it is anticipated that many of the methods now classified as advanced will become conventional within the next 5 to 10 years.

Compounds containing available nitrogen and phosphorus have received considerable attention since the mid-1960s. Initially, nitrogen and phosphorus in wastewater discharges became important because of their effects in accelerating eutrophication of lakes and promoting aquatic growth. More recently, nutrient control has become a routine part of treating wastewaters used for the recharge of groundwater supplies. Nitrification of wastewater discharges is also require in many cases to reduce ammonia toxicity or to lessen the impact on the oxygen resources in flowing streams or estuaries. As a result of the many concerns over nutrients, nutrient removal has become, for all practical purposes, an integral part of conventional wastewater treatment.

Technologies Used for Advanced Treatment

Advanced wastewater treatment systems may be classified by the type of unit operation or process or by the principal removal function performed. To facilitate a general comparison of the various operations and processes, information on 1) the principle residual constituent removal function and 2) the types of operations or processes that can be used to perform this function are presented. The processes listed in Tab. 25.1 can be grouped into four broad categories requiring removal: 1) residual organic colloidal and suspended solids, 2) dissolved organic constituents, 3) dissolved inorganic constituents, and 4) biological constituents.

Tab. 25.1 Typical process flow diagrams for wastewater treatment employing advanced treatment process with settled secondary effluent

	Process flow diagrams for wastewater treatment employing advanced treatment
1	secondary effluent→coagulation→filtration→carbon adsorption→Cl_2 or UV disinfection→
2	secondary effluent→coagulation→flocculation→filtration →Cl_2 or UV disinfection→
3	secondary effluent→coagulation→flocculation→sedimentation→filtration →Cl_2 or UV disinfection→
4	secondary effluent→ lime precipitation→ recarbonation→ filtration carbon adsorption→ ozonation→ UV disinfection→Cl_2 disinfection→
5	secondary effluent→ultrafiltration→reverse osmosis→ozonation→UV disinfection→Cl_2 disinfection→
6	secondary effluent→microfiltration→reverse osmosis→ion exchange→ozonation→Cl_2 disinfection→
7	secondary effluent→microfiltration→reverse osmosis→UV disinfection→Cl_2 disinfection→

Selection of a given operation, process, or combination therefore depends on 1) the use to be made of the treated effluent, 2) the nature of the wastewater, 3) the compatibility of the various operations and processes, 4) the available means to dispose of the ultimate contaminants, and 5) the environmental and economic feasibility of the various systems. It should be noted that in some situations economic feasibility may not be a controlling factor in the design of advanced wastewater treatment systems, especially where specific constituents must be removed to protect the environment. Based on the variations in performance observed in the field, pilot plant testing is recommended for the development of treatment performance data and design criteria. Representative performance data are presented in the discussion of the individual technologies.

Vocabulary

colloidal	胶体的	eutrophication	富营养化
potassium	钾	ultrafiltration	超滤
stringent	严格的	reverse osmosis	反渗透
residual	残留的		

Reading Material A

An Introduction of Depth Filtration

Depth filtration involves the removal of particulate material suspended in a liquid by passing the liquid through a filter bed comprised of a granular or compressible filter medium. Although depth filtration is one of the principal unit operations used in the treatment of potable water, the filtration of effluents from wastewater treatment process is becoming more common. Depth filtration is now used to achieve supplemental removals of suspended solids (including particulate BOD) from wastewater effluents of biological and chemical treatment processes to reduce the mass discharge of solids and, perhaps more importantly, as a conditioning step that will allow for the effective disinfection of the filtered effluent. Depth filtrations also used as a pretreatment step for membrane filtration. Single-and two-stage filtration is also used to remove chemically precipitated phosphorus.

Historically, the first depth filtration process developed for the treatment of wastewater was the slow sand filter (typical filtration rates of 30 to 60 $L/(m^2 \cdot d)$, Frankland (1870), and Dunbar (1908)). The rapid sand filter (typical filtration rates of 80 to 200 $L/(m^2 \cdot min)$, the subject of this section, was developed to treat larger volumes of water in a facility with a smaller footprint. To introduce the subject of depth filtration, the purpose of this section is to present 1) a general introduction to the depth filtration process, 2) an introduction to filter clean-water hydraulics, and 3) an analysis of the filtration process. The types of filters that are available and issues associated with their selection and design, including a discussion of the need for pilot-plant studies, are considered in the following section.

Before discussing the available filter technologies, it will be useful to first describe the basics of the depth filtration including 1) the physical features of a conventional granular medium-depth filter, 2) filter-medium characteristics, 3) the filtration process in which suspended material is removed from the liquid, 4) the operative particle-removal mechanisms that bring about the removal of suspended material within the filter, and 5) the backwash process, in which the material that has been retained within the filter is removed.

Physical Features of a Depth Filter

The general features of a convention rapid granular medium-depth filter are illustrated on Fig. 25.1. As shown, the filtering medium (sand in this case) is supported on a gravel layer, which, in turn, rests on the filter under-drain system. The water to be filtered enters the filter from an inlet channel. Filtered water is collected in the underdrain system, which is also used to reverse the flow to backwash the filter. Filtered water typically is disinfected before being discharged to the environment. If the filtered water is to be reused, it can be discharged to a storage reservoir or to the reclaimed water distribution system. The hydraulic control of the filter is described in a subsequent section.

Filter-medium Characteristics

Grain size is the principal filter-medium characteristic that affects the filtration operation. Grain size affects both the clear-water headloss and the buildup of headloss during the filter run. If too small a filtering medium is selected, much of the driving force will be wasted in overcoming the frictional resistance of the filter bed. On the other hand, if the size of the medium is too large, many of the small particles in the influent will pass directly through the bed.

The Filtration Process

During filtration in a conventional downflow depth filter, wastewater containing suspended matter is applied to the top of the filter bed (Fig. 25.1(a)). As the water passes through the filter bed, the suspended matter in the wastewater is removed by a variety of removal mechanisms as described below. With the passage of time, as material accumulates within the interstices of the granular medium, the headloss through the filter starts to build up beyond the initial value, as shown in Fig. 25.1. After some period of time, the operating headloss or effluent turbidity reaches a predetermined headloss or turbidity value, and the filter must be cleaned. Under ideal conditions, the time required for the headloss buildup to reach the preselected terminal value should correspond to the time when the suspended solids in the effluent reach the preselected terminal value for acceptable quality. In actual practice, one or the other event will govern the backwash cycle.

Particle Removal Mechanisms

Straining has been identified as the principal mechanism that is operative in the removal of suspended solids during the filtration of settled secondary effluent from biological treatment process (Tchobanoglous and Eliassen, 1970). Other mechanisms including impaction, interception, and adhesion are also operative even though their effects are small and, for the most part, masked by the straining action.

The removal of the smaller particles found in wastewater must be accomplished in two steps involving 1) the transport of the particles to or near the surface where they will be removed and 2) the

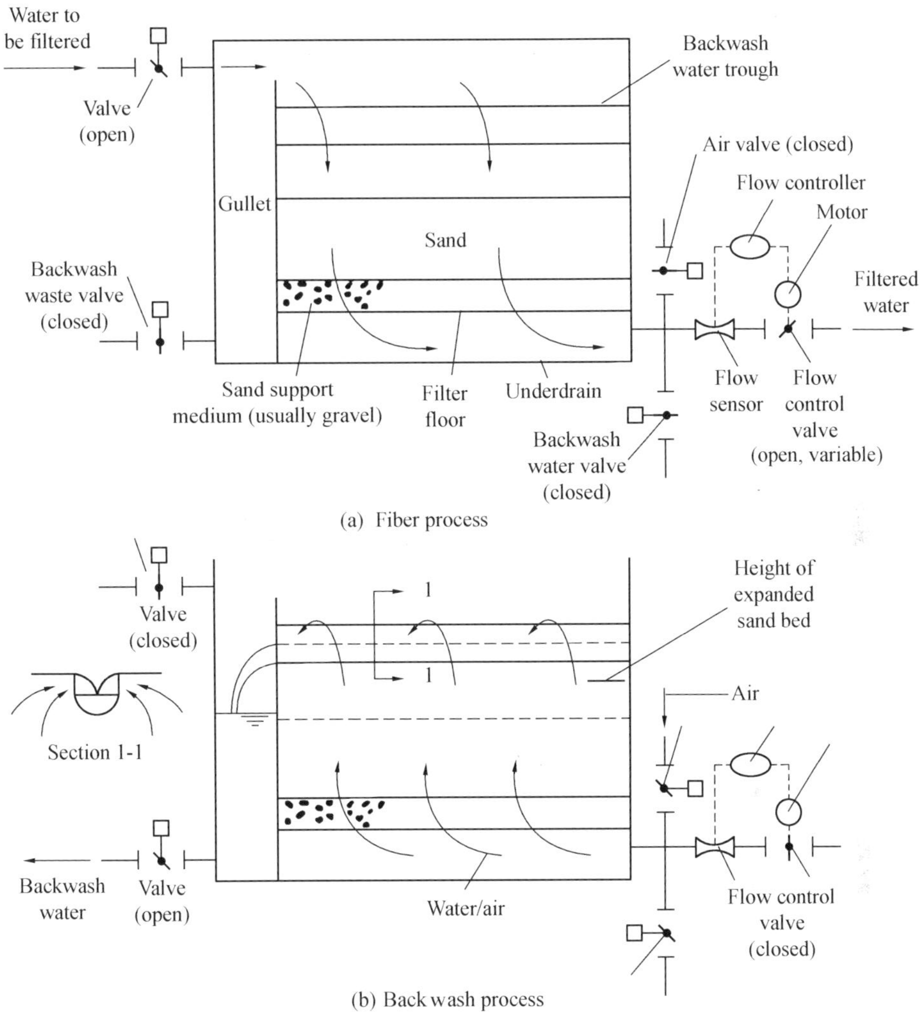

Fig. 25.1 General features and operation of a conventional rapid granular medium-depth filter

removal of particles by one or more of the operative removal mechanisms. This two-step process has been identified as transport and attachment (O'Melia and Stumm, 1967).

Backwash Process

The end of the filter run (filtration phase) is reached when the suspended solids in the effluent start to increase (break through) beyond an acceptable level, or when a limiting headloss occurs across the filter bed. Once either of these conditions is reached, the filtration phase is terminated, and the filter must be cleaned (backwashed) to remove the material (suspended solids) that has accumulated within the granular filter bed. Backwashing is accomplished by reversing the flow through the filter. A sufficient flow of washwater is applied until the granular filter medium is fluidized (expanded), causing the particles of the filtering medium to abrade against each other.

The suspended mater arrested within the filter is removed by the shear forces created by backwash water as it moves up through the expanded bed. The material that has accumulated within the bed is then washed away. Surface washing with water and air scouring are often used in conjunction with the water backwash to enhance the cleaning of the filter bed. In most wastewater treatment plant flow diagrams, the washwater containing the suspended solids that are removed from the filter is returned either to the primary settling facilities or to the biological treatment process.

Vocabulary

depth filtration	深床过滤	preselected	预先挑选的，预先设定的
membrane filtration	膜过滤		
filtering medium	过滤介质	backwash	反冲洗

Reading Material B

Membrane Filtration Processes

Filtration involves the separation (removal) of particulate and colloidal matter from a liquid. In membrane filtration the range of particle size is extended to include dissolved constituents (typically 0.000 1 to 1.0 μm). The role of membrane is to serve as a selective barrier that will allow the passage of certain constituents and will retain other constituent found in the liquid (Cheryan, 1998). To introduce membrane technologies and their application, the following subjects are considered in this section: 1) membrane process terminology, 2) membrane classification, 3) membrane configurations, and 4) application of membrane technologies.

Membrane Process Terminology

Terms will be commonly encountered when considering the application of membrane process. The influent to the membrane module is known as the feed stream (also known as feedwater). The liquid that passes throught the semipermeable membrane is known as permeate (also known as the product stream or permeating stream) and the liquid containing the retain constituents is known as the concentrate (also known as the retentate, reject, retain phase or waster stream). The rate at which the permeate flows through the membrane is known as the rate of flux, typically expressed as kg/(m·d).

Membrane Process Classification

Membrane process include microfiltratin (MF), ultrafiltration (UF), nanofiltration (NF), reverse osmosis (OS), dialysis, and electrodialysis (ED), membrane processes can be classified in a number of different ways including 1) the type of material from which the membrane is made 2) the nature of the driving force 3) the separation mechanism and 4) the nominal size of the separation achieved.

Membranes Materials

Membranes used for the treatment of water and wastewater typically consist of a thin skin having a thickness of about 0.2 to 0.25 μm supported by a more porous structure of about 100 μm in thickness. Most commercial membrane are produced as flat sheets, fine hollow fibers, or in tubular form. The flat sheets are of two types, asymmetric and composite. Asymmetric membranes are cast in one process and consist of a very thin (less than 1 μm) layer and a thicker (up to 100 μm) porous layer that adds support and is capable of high water flux. Thin-film composite (TFC) membrane are made by bonding a thin cellulose acetate, polyamide, or other active layer (typically 0.15 to 0.25 μm thick) to a thick porous substrate, which provides stability. Membranes can be made from a number of different organic and inorganic materials. The membranes used for wastewater treatment are typically organic. The principal types of membranes used include polypropylene, cellulose acetate, aromatic polyamides, and thin-film composite (TFC). The choice of membrane and system configuration is based on minimizing membrane clogging and deterioration, typically based on pilot-plant studies.

Membrane Systems

The main force of membrane technology is the fact that it works without the addition of chemicals, with a relatively low energy use and easy and well-arranged process conductions. Membrane technology is a generic term for a number of different, very characteristic separation processes. These processes are of the same kind, because in each of them a membrane is used. Membranes are used more and more often for the creation of process water from groundwater, surface water or wastewater. Membranes are now competitive for conventional techniques. The membrane separation process is based on the presence of semi permeable membranes. The principle is quite simple: the membrane acts as a very specific filter that will let water flow through, while it catches suspended solids and other substances. There are various methods to enable substances to penetrate a membrane. Examples of these methods are the applications of high pressure, the maintenance of a concentration gradient on both sides of the membrane and the introduction of an electric potential. Membranes occupy through a selective separation wall. Certain substances can pass through the membrane, while other substances are caught. Membrane filtration can be used as an alternative for flocculation, sediment purification techniques, adsorption (sand filters and active carbon filters, ion exchangers), extraction and distillation.

There are two factors that determine the affectivity of a membrane filtration process: selectivity and productivity. Selectivity is expressed as a parameter called retention or separation factor (expressed by the unit $L/(m^2 \cdot h)$). Productivity is expressed as a parameter called flux (expressed by the unit $L/(m^2 \cdot h)$). Selectivity and productivity are membrane-dependent.

When membrane filtration is used for the removal of larger particles, microflitration and ultra filtration are applied. Because of the open character of the membranes the productivity is high while the pressure differences are low. When salts need to be removed from water, nano filtration and Reverse Osmosis are applied. Nanofiltration and RO membranes do not work according to the principle of pores. Separation takes place by diffusion through the membrane. The pressure that is required to perform nano filtration and Reverse Osmosis is much higher than the pressure required for micro and ultra filtration, while productivity is much lower.

The choice for a certain kind of membrane system is determined by a great number of aspects, such as costs, risks of plugging of the membranes, packing density and cleaning opportunities. Membranes are never applied as one flat plate, because this large surface often results in high investing costs. That is why systems are built densely to enable a large membrane surface to be put in the smallest possible volume. Membranes are implemented in several types of modules. There are two main types, called the tubular membrane system and the plate & frame membrane system. Tubular membrane systems are divided up in tubular, capillary and hollow fiber membranes. Plate & frame membranes are divided up in spiral membranes and pillow-shaped membranes.

During membrane filtration processes membrane fouling is inevitable, even with a sufficient pre-treatment. The types and amounts of fouling are dependent on many different factors, such as feed water quality, membrane type, membrane materials and process design and control. Particles, biofouling and scaling are the three main types of fouling on a membrane. These contaminants cause that a higher workload is required, to be able to guarantee a continuous capacity of the membranes. At a certain point the pressure will rise so much that it is no longer economically and technically accountable. There are a number of cleaning techniques for the removal of membrane fouling. These techniques are forward flushing, backward flushing, air flushing and chemical cleaning, and any combination of the methods.

Vocabulary

terminology	术语	asymmetric	不均匀的
semipermeable	半透性的	electrodialysis	电渗析
feedwater	给水,进水	nanofiltration	纳滤
microflitration	微滤	polyamide	聚酰胺

Unit 26　Adsorption

Adsorption is the process of accumulating substances that are in solution on a suitable interface. Adsorption is a mass transfer operation in which a constituent in the liquid phase is transferred to the solid phase. The adsorbate is the substance that is being removed from the liquid phase at the interface. The adsorbent is the solid, liquid, or gas phase onto which the adsorbate accumulates. Although adsorption is used at the air-liquid interface in the flotation process, only the case of adsorption at the liquid-solid interface will be considered in this discussion. The adsorption process has not been used extensively in wastewater treatment, but demands for a better quality of treated wastewater effluent, including toxicity reduction, have led to an intensive examination and use of the process of adsorption on activated carbon. Activated carbon treatment of wastewater is usually thought of as a polishing process for water that has already received normal biological treatment. The carbon in this case is used to remove a portion of the remaining dissolved organic matter. The purpose of this section is to introduce the basic concepts of adsorption and to consider carbon adsorption.

The principal types of adsorbents include activated carbon, synthetic polymeric, and silica-based adsorbents, although synthetic polymeric and silica-based adsorbents are seldom used for wastewater adsorption because of their high cost. Because activated carbon is used most commonly in advanced wastewater-treatment applications, the focus of the following discussion is on activated carbon. The nature of activated carbon, the use of granular carbon and powdered carbon for wastewater treatment, and carbon regeneration and reactivation are discussed below.

Activated Carbon

Activated carbon is prepared by first making a char from organic materials such as almond, coconut, and walnut hulls, other materials including woods, bone, and coal have also been used. The char is produced by heating the base material to a red heat(less than about 700 ℃) in a retort to drive off the hydrocarbons, but with an insufficient supply of oxygen to sustain combustion. The carbonization or char-producing process is essentially a pyrolysis process. The char particle is then activated by exposure to oxidizing gases such as steam and CO_2 at high temperatures, in the range from 800 to 900 ℃. These gases develop a porous structure in the char and create a large internal surface area.

The surface properties that result are a function of both the initial material used and the preparation procedure, so that many variations are possible. The type of base material from which the activated carbon is derived may also affect the pore-size distribution and the regeneration characteristics. After activation, the carbon can be separated into, or prepared in, different sizes with different adsorption capacity. The two size classifications are powdered activated carbon (PAC), which typically has a diameter of less than 0.074 mm, and granular activated carbon (GAC), which has a diameter greater than 0.1 mm.

Carbon Regeneration and Reactivation

Economical application of activated carbon depends on an efficient means of regenerating and reactivating the carbon after its adsorptive capacity has been reached. Regeneration is the term used to describe all of the processes that are used to recover the adsorptive capacity of the spent carbon, exclusive of reactivation, including: 1) chemicals to oxidize the adsorbed material, 2) steam to drive off the adsorbed material, 3) solvents, and 4) biological conversion processes. Typically some of the adsorptive capacity of the carbon (about 4 to 10 percent) is also lost in the regeneration method used (Crittenden, 2000). In some applications, the capacity of the carbon following regeneration has remained essentially the same for years. A major problem with the use of powdered activated carbon is that the methodology for its regeneration is not well defined. The use of powdered activated carbon produced from recycled solid wastes may obviate the need to regenerate the spent carbon, and may be more economical.

Reactivation of granular carbon involves essentially the same process used to create the activated carbon from virgin material. Spent carbon is reactivated in a furnace by oxidizing the adsorbed organic material and, thus, removing it from the carbon surface. The following series of events occurs in the reactivation of spent activated carbon: 1) the carbon is heated to drive off the adsorbed organic material (i.e., absorbate), 2) in the process of driving off the adsorbed material some new compounds are formed that remain on the surface of the carbon, and 3) the final step in the reactivation process is to burn off the new compounds that were formed when the absorbed material was to burn off. With effective process control, the adsorptive capacity of reactivated carbon will be essentially the same as that of the virgin carbon (Crittenden, 2000). For planning purposes, it is often assumed that a loss of 2 to 5 percent will occur in the reactivation process. It is important to note that most other losses of carbon occur through attrition due to mishandling. For example, right angle bends in piping cause attrition through abrasion and impact. The type of pumping facilities used will also affect the amount of attrition. In general, a 4 to 8 percent loss of carbon is assumed, due to handling. Replacement carbon must be available to make up the loss.

Fundamentals of Adsorption

The adsorption process takes place in four more or less definable steps: 1) bulk solution transport, 2) film diffusion transport, 3) pore transport, and 4) adsorption (or sorption). Bulk solution transport involves the movement of the organic material to be adsorbed through the bulk liquid to the boundary layer of fixed film of liquid surrounding the adsorbent, typically by advection and dispersion in carbon contactors. Film diffusion transport involves the transport by diffusion of the organic material through the stagnant liquid film to the entrance of the pores of the adsorbent. Pore transport involves the transport of the material to be adsorbed through the pores by a combination of molecular diffusion through the pore liquid and/or by diffusion along the surface of the adsorbent. Adsorption involves the attachment of the material to be adsorbed to adsorbent at an available adsorption site (Snoeyink and Summers, 1999). Adsorption can occur on the outer surface of the adsorbent and in the macropores, mesopores, micropores, and submicropores, but the surface area of the macro-and mesopores is small compared with the surface area of the micropores and submicropores and the amount of material adsorbed there is usually considered negligible. Adsorption forces include (Crittenden, 1999): 1) Coulombic-unlike charges; 2) Point charge and a dipole; 3) Dipole-dipole interactions; 4) Point charge neutral species; 5) London or van der Waals forces; 6) Covalent bounding with reaction; 7) Hydrogen bonding.

Additional details on the above adsorption force may be found in Crittenden (1992). Because it is difficult to differentiate between chemical and physical adsorption, the term sorption is often used to describe the attachment of the organic material to the activated carbon.

Because the adsorption process occurs in a series of steps, the slowest step in the series is identified as the rate limiting step. In general, if physical adsorption is the principal method of adsorption, one of the diffusion transport steps will be the rate limiting. When the rate of sorption equals the rate of desorption, equilibrium has been achieved and the capacity of the carbon has been reached. The theoretical adsorption capacity of the carbon for a particular contaminant can be determined by developing its adsorption isotherm as described below.

Vocabulary

adsorption	吸附	furnace	炉子,熔炉
adsorbent	吸附剂	obviate	排除,避免
adsorbate	被吸附物	methodology	方法学,方法论
regeneration	再生	macropore	大孔,大孔隙
almond	杏树	mesopore	中孔,中孔隙
coconut	椰子	micropore	微孔,微孔隙
walnut hull	胡桃壳	desorption	解吸
pyrolysis	高温分解		

Reading Material A

Ion Exchange

Ion exchange is a unit process in which ions of a given species are displaced from an insoluble exchange material by ions of a different species in solution. The most widespread use of this process is in domestic water softening, where sodium ions from a cationic-exchange resin replace the calcium and magnesium ions in the treated water, thus reducing the hardness. Ion exchange has been used in wastewater applications for the removal of nitrogen, heavy metals, and total dissolved solids.

Ion exchange processes can be operated in a batch or continuous mode. In a batch process, the resin is stirred with the water to be treated in a reactor until the reaction is complete. The spent resin is removed by settling and subsequently is regenerated and reused. In a continuous process, the exchange material is placed in a bed or a packed column, and the water to be treated is passed through it. Continuous ion exchangers are usually of the downflow, packed-bed column type. Wastewater enters the top of the column under pressure, passes downward through the resin bed, and is removed at the bottom. When the resin capacity is exhausted, the column is backwashed to remove trapped solids and is then regenerated.

Ion-exchange Materials

Naturally occurring ion-exchange materials, known as zeolites, are used for water softening and ammonium ion removal. Zeolites used for water softening are complex aluminosilicates with sodium as the mobile ion. Ammonium exchange is accomplished using a naturally occurring zeolite clinoptilolite. Synthetic aluminosilicates are manufactured, but most synthetic ion-exchange materials are resins or phenolic polymers. Five types of synthetic ion-exchange resins are in use: 1) strong-acid cation, 2) weak-acid cation, 3) strong-base anion, 4) weak-base anion, and 5) heavy-metal selective chelating resins. The properties of these resins are summarized in Tab. 26.1.

Tab. 26.1 Classification of exchange resin

Type of resin	Characteristics
Strong-acid cation resins	Strong-acid resins behave in a manner similar to a strong acid, and are highly ionized in both the acid and salt form over the entire pH range.
Weak-acid cation resins	Weak-acid cation exchangers have a weak-acid functional group, typically a carboxylic group. These resins behave like weak organic acids that are weakly dissociated
Strong-base anion resins	Strong-base resins are highly ionized, having strong-base functional groups such as, and can be used over the entire pH range. These resins are used in the hydroxide form for water deionization
Weak-base anion resins	Weak-base resins have weak-base functional groups in which the degree of ionization is dependent on pH
Heavy-metal selective chelating resins	Chelating resins behave like weak-acid cation resins but exhibit a high degree of selectivity for heavy-metal cations. The functional group in most of these resins is EDTA, and the resin structure in the sodium form is R-EDTA-Na

Most synthetic ion-exchange resins are manufactured by a process in which styrene and divinylbenzene are copolymerized. The styrene serves as the basic matrix of the resin, and divinylbenzene is used to cross-link the polymers to produce an insoluble tough resin. Important properties of ion-exchange resins include exchange capacity, particle size, and stability. The exchange capacity of a resin is defined as the quantity of an exchangeable ion that can be taken up. The exchange capacity of resins is expressed as eq/kg or eq/L. The particle size of a resin is important with respect to the hydraulics of the ion-exchange column and the kinetics of ion exchange. In general, the rate of exchange is proportional to the inverse of the square of the particle diameter. The stability of a resin is important to the long-term performance of the resin. Excessive osmotic swelling and shrinking, chemical degradation, and structural changes in the resin caused by physical stresses are important factors that may limit the useful life of a resin.

Typical Ion-exchange Reactions

Typical ion-exchange reactions for natural and synthetic ion-exchange material are given below.

For natural zeolites (Z):

$$ZNa_2 + \begin{bmatrix} Ca^{2+} \\ Mg^{2+} \\ Fe^{2+} \end{bmatrix} \leftrightarrow Z\begin{bmatrix} Ca^{2+} \\ Mg^{2+} \\ Fe^{2+} \end{bmatrix} + 2Na$$

For synthetic resins(R):

Strong-acid cation exchange:

$$RSO_3H + Na^+ \leftrightarrow RSO_3Na + H^+$$

$$2RSO_3Na + Ca^{2+} \leftrightarrow (RSO_3)_2Ca + 2Na^+$$

Weak-acid cation exchange:

$$RCOOH + Na^+ \leftrightarrow RCOONa + H^+$$

$$2RCOOHNa + Ca^{2+} \leftrightarrow (RCOO)_2Ca + 2Na^+$$

Strong-base anion exchange:

$$RR'_3NOH + Cl^- \leftrightarrow RR'_3NCl + OH^-$$

Weak-base anion exchange:

$$RNH_3 + Cl^- \leftrightarrow RNH_3Cl + OH^-$$

$$2RNH_3Cl + SO_4^{2-} \leftrightarrow (RNH_3)_2SO_4 + 2Cl^-$$

Vocabulary

ion exchange	离子交换	aluminosilicate	铝矽酸盐
resin	树脂	phenolic	酚的，石碳酸的
cation	阳离子	deionization	离子化,电离
zeolite	沸石	divinylbenzene	二乙烯基苯
zeolite clinoptilolite	斜发沸石	copolymerize	(使)共聚合

Reading Material B

Advanced Wastewater Treatment

Although secondary treatment processes, when coupled with disinfection, may remove over 85 percent of the BOD and suspended solids and nearly all pathogens, only minor removal of some pollutants, such as nitrogen, phosphorus, soluble COD, and heave metals, is achieved. In some circumstances, these pollutants may be of major concern. In these cases, processes capable of removing pollutants not adequately removed by secondary treatment are used in what is called advanced wastewater treatment (these processes have often been called advanced wastewater treatment, or AWT for short).

The following sections describe available AWT processes. In addition to solving tough pollution problems, these processes improve the effluent quality to point that it is adequate for many reuse purposes, and may convert what was originally a wastewater into a valuable resource too good to throw away.

Phosphorus Removal

All the polyphosphates (molecularly dehydrated phosphates) gradually hydrolyze in aqueous solution revert to the ortho form (PO_4^{3-}) from which they were derived. Phosphorus is typically found as mono-hydrolyze phosphate (HPO_4^{2-}) in wastewater. The removal of phosphorus to prevent or reduce eutrophication is typically accomplished by chemical precipitation using one of three compounds. The precipitation reactions for each are shown below.

Using ferric chloride

$$FeCl_3 + HPO_4^{2-} \Leftrightarrow FePO_4 + H^+ + 3Cl^-$$

Using alum

$$Al_2(SO_4)_3 + 2HPO_4^{2-} \Leftrightarrow 2AlPO_4 \downarrow + 2H^+ + 3SO_4^{2-}$$

Using lime

$$5Ca(OH)_2 + 3HPO_4^{2-} \Leftrightarrow Ca_5(PO_4)_3OH \downarrow + 3H_2O + 6OH^-$$

You should note that ferric chloride and alum reduce the pH while lime increases it. The effective range of pH for alum and ferric chloride is between 5.5 and 7.0. If there is not enough naturally occurring alkalinity to buffer the system to this range, then lime must be added to counteract the formation of H^+.

The precipitation of phosphorus requires a reaction basin and a setting tank to remove the precipitate. When ferric chloride and alum are used, the chemicals may be added directly to the aeration tank in the activated sludge system. Thus, the aeration tank serves as a reaction basin. The precipitate is then removed in the secondary clarifier. This is not possible with lime since the high pH required to form the precipitate is determined to the activated sludge organisms. In some wastewater treatment plant, the $FeCl_3$ (or alum) is added before the wastewater enters the primary sedimentation tank. This improves the efficiency of the primary tank, but deprive the biological processes of needed nutrients.

Nitrogen Control

Nitrogen in any soluble form (NH_3, NH_4^+, NO_2^-, and NO_3^-, but not N_2 gas) is a nutrient and may need to be removed from wastewater to help control algae in the receiving body. In addition, nitrogen in the form of ammonia exerts an oxygen demand and can be toxic to fish. Removal of nitrogen can be accomplished either biologically or chemically. The biological process is called nitrification/denitrification. The chemical process is called ammonia stripping.

Nitrification/Denitrification The natural nitrification processes can be forced to occur in the activated sludge system by maintaining a cell detention time of 15 days or more. The nitrification step is expressed in chemical terms as follows:

$$NH_4^+ + 2O_2 = NO_3^- + H_2O + 2H^+$$

Of course, bacteria must be present to cause the reaction to occur. This step satisfies the oxygen demand of the ammonium ion. If the nitrogen level is not of concern for the receiving body, the wastewater can be discharged after setting.

If nitrogen is of concern, the nitrogen step must be followed by anoxic denitrification by bacteria:

$$2NO_3^- + \text{organic matter} \longrightarrow N_2 + CO_2 + H_2O$$

As indicated by the chemical reaction, organic matter is required for denitrification. Organic matter serve as an energy source for the bacteria. The organics may be obtained from within or outside the cell. In the multistage nitrogen removal systems, because the concentration of BOD_5 in the flow to the denitrification process is usually quite low, a supplemental organic carbon is required for rapid denitrification (BOD_5 concentration is low because the wastewater previously has undergone carbonaceous BOD_5 removal and nitrification process). The organic matter may be either raw, settled sewage or a synthetic material such as methanol (CH_3OH). Raw, settled sewage may adversely affect the effluent quality by increasing the BOD_5 and ammonia content.

Ammonia Stripping Nitrogen in the form of ammonia can be removed chemically from water by raising the pH to ammonium ion into ammonia, which can be stripped from the water by passing large quantities of through the water. The process has no effect on nitrate, so the activated sludge process must be operated at a short cell detention time to prevent nitrification. The ammonia stripping reaction is $NH_4^+ + OH^- \Leftrightarrow NH_3 + H_2O$.

The hydroxide is usually supplied by adding lime. The lime also reacts with CO_2 in the air and water to form a calcium carbonate scale, which must be removed periodically. Low temperatures cause problems with icing and reduced stripping ability. The reduced stripping ability is caused by the increased solubility of ammonia in cold water.

Besides chemical precipitation of phosphorous, biological nitrification-denitrification, there are filtration process similar to that used in water treatment plants, carbon adsorption and biological phosphorous removal, those are also important components of advanced wastewater treatment.

Vocabulary

polyphosphoric acid	多磷酸	nitrification	硝化,硝化作用
ortho	正,原	denitrification	反硝化,反硝化作用
eutrophication	富营养化	anoxic	缺氧的
hydrolyze	水解	ammonia stripping	氨吹脱
counteract	减少,抵抗,反作用		

Unit 27 Anaerobic Fermentation

Anaerobic fermentation and oxidation processes are used primarily for the treatment of waste sludge and high-strength organic wastes. However, application for dilute waste streams have also been demonstrated and are becoming more common. Anaerobic fermentation processes are advantageous because of the lower biomass yields and because energy, in the form of methane, can be recovered from the biological conversion of organic substrates. Although most fermentation processes are operated in the mesophilic temperature range (30 to 35 ℃), there is increased interest in thermophilic fermentation along or before mesophilic fermentation. The latter is termed temperature phased anaerobic digestion (TPAD) and is typically designed with a sludge SRT of 3 to 7 d in the first thermophilic phase at 50 to 60 ℃ and 7 to 15 d in the final mesophilic phase (Han and dague, 1997). Thermophilic anaerobic digestion processes, are used to accomplish high pathogen kill to produce Class A biosolids, which can be used for unrestricted reuse applications.

For treating high-strength industrial wastewaters, anaerobic treatment has been shown to provide a very cost-effective alternative to aerobic processes with saving in energy, nutrient addition, and reactor volume. Because the effluent quality is not as good as that obtained with aerobic treatment, anaerobic treatment is commonly used as a pretreatment step prior to discharge to a municipal collection system or is followed by an aerobic process.

Anaerobic treatment converts organic wastes to methane and carbon dioxide in the absence of air. Anaerobic decomposition is a three stage reaction: 1) hydrolysis of suspended organic solids into soluble organic compounds; 2) acidogenesis, or conversion of soluble organics to volatile fatty acids; and 3) methanogenesis, or conversion of the volatile fatty acids into methane. Hydrolysis does not occur in all anaerobic treatment systems. However, for certain wastes, such as municipal wastewater treatment sludges, waste pharmaceutical cells, and some food waste hydrolysis is an important reaction.

In some cases, the hydrolysis step can be the rate limiting reaction. The second step, acidogenesis, is accomplished by a general class of microorganisms known as the volatile acid formers. The products of the acid forming reaction are the acid salts of short-chain fatty acids, primarily acetic, propionic, and butyric acid. Other products of the acid-forming reaction are carbon dioxide and new bacterial cells. The acid-forming steps can proceed over a broad range of environmental conditions, such as pH and temperature.

The third and final step in the series of anaerobic digestion reactions is the formation of methane and carbon dioxide. For most organic wastes, methane and carbon dioxide are formed from the decomposition of acetic acid. Methane, however, can be produced by decomposition of propionic and other volatile acids. The methane-forming bacteria are more sensitive to environmental conditions, such as pH, temperature, and inhibitory compounds, and are more fragile than the acid-forming bacteria. Therefore, it takes more time for the methane bacteria to recover from inhibition or shock conditions than it does for the acid-forming bacteria.

Anaerobic treatment can be defined biochemically as the conversion of organic compounds into carbon dioxide, methane, and microbial cells. For anaerobic treatment of organic nitrogen compounds such as domestic sludge, the end products will also include ammonia. The portion of the organic compound that is termed the net cell yield. The cell yield will range from approximately 5 percent for fatty acids to approximately 20 percent for carbohydrates. Due to low yield coefficients, methane and carbon dioxide production account for the majority of the influent organics, i. e., COD, processed. The theoretical methane yield, with a bacterial yield coefficient of zero, is 0. 370 m^3/kg COD degraded. Since carbon dioxide is partially soluble in the wastewater, the gas production will be dependent on pH and alkalinity. Gas production rates applied to the influent COD must also consider the nonbiodegradable fraction of the COD. Since COD reduction is stoichiometricallu related to methane production, the anaerobic digester performance can be easily calculated.

pH and Alkalinity Requirement

The optimum pH range for methane fermentation is between 7and 8, but the pH is and the longer the pH is at a low value, the more likely is a damaging upset. For this reason, it is recommended that the pH be maintained above 6.5 for anaerobic reactors. When the bicarbonate alkalinity drops below about 500 mg/L, and with the normal percentage of carbon dioxide (approximately 38 percent) in the reactor gas, the pH will drop close to 6.0. Anaerobic treatment of organic nitrogen-condition wastes produces ammonium and bicarbonate which create alkalinity for the treatment system. The ammonium and bicarbonate which create alkalinity for the treatment system. The ammonium cation (NH_4^+) is in equilibrium with bicarbonate, without ammonium in solution, carbon dioxide would leave the solution as carbonate ion.

Nutrient Requirements

Anaerobic microorganisms require nutrients to sustain growth. The nutrient requirements for anaerobic microorganisms are different from aerobic requirements because the cell yield is lower. Furthermore, the nitrogen and phosphorus requirements for anaerobic treatment are typically only 20 percent of those for aerobic treatment. Whereas a COD:N:P ratio of 100:5:1 is typical of aerobic treatment, a COD:N:P ratio of 100:1:0.2 is typical of anaerobic growth. Specific nitrogen and phosphorus requirements will depend upon the nature of the organic compounds to be biodegraded and the sludge age of the treatment system.

In addition to nitrogen and phosphorus, anaerobic microorganisms require other nutrients for growth. These nutrient (sulfur, iron, calcium, magnesium, sodium, and potassium) are typically present in sufficient concentrations in many industrial wastes. However, for some wastes, they can be deficient and may need to be added.

Temperature

As with other biological processes, anaerobic bacteria can be divided into three temperature ranges. In general, the higher the temperature of the anaerobic system is, the faster the substrate removal rate and the faster the cell decay rate. Normally, anaerobic reactors are operated in the mesophilic range, 25 to 40 ℃, or the thermophilic range, 50 to 70 ℃. Thermophilic systems will require a smaller reactor than a mesophilic system. However, because of their low cell yield, thermophilic processes are very slow to start up and cannot tolerate variations in loading, variations in substrate characteristics, or toxic compounds.

Toxicity

Methanogens are commonly considered to be the most sensitive to toxicity of all the microorganisms involved in the anaerobic treatment of organic wastewaters. However, methane bacteria, like most microorganisms, can tolerate a wide variety of toxicants. In addition, many toxic compounds are biodegraded in anaerobic reactors so that the methanogens are not affected by them. Acclimation to toxic compounds and reversibility of toxic effects are known to occur. The toxicity of a compound is related to its concentration, the length of exposure time, and the normal background concentration. The toxicity and/or inhibition actions of some common cations are stimulatory at low concentrations but are toxic at higher concentrations. Toxicity can be prevented by addition of other cations which act as cation antagonists. For example, the toxic effect of sodium can be reduced by adding potassium and further reduced by adding calcium. Antagonists are best added as a chloride salt. If antagonists are not available or are too costly, the best method to prevent toxicity may be dilution.

Hydrogen sulfide has been recognized to be toxic to anaerobic microorganism, especially methanegens, under strictly anaerobic conditions, the sulfate ion (SO_4^{2-}) is biochemically reduced to the sulfide ion (S^{2-}) which establishes an equilibrium with the various forms of hydrogen sulfide (H_2S, HS^- and S^{2-}). The toxic concentration of total dissolved sulfide in anaerobic digestion has been reported to be 200 to 300 mg/L.

Vocabulary

fermentation	发酵	methanogenesis	产甲烷
mesophilic	中温的,(细菌)嗜温的	propionic	丙酸
thermophilic	高温的,嗜热的	fragile	脆弱的,虚弱的
pharmaceutical	药物的,药剂的	stoichiometricallu	化学计量学的
hydrolysis	水解	antagonist	对手,对抗药
acidogenesis	酸化,产酸		

Reading Material A

Wastewater Reuse

In many locations where the available supply of fresh water has become inadequate to meet water needs, it is clear that the once-used water collected from communities and municipalities must be viewed not as a waste to be disposed of but as a resource that must be reused. The concept of reuse is becoming accepted more widely as other parts of the country experience water shortages. The use of dual water systems, such as now used in St. Petersburg in Florida and Rancho Viejo in California, is expected to increase in the future. In both locations, treated effluent is used for landscape watering and other nonpotable uses. Satellite reclaimation systems such as those used in the Los Angeles basin, where wastewater flows are mined (withdrawn from collection systems) for local treatment and reuse, are examples where transportation and treatment costs of reclaimed water can be reduced significantly. Because water reuse is expected to become of even greater importance in the future.

Current Status

Most of the reuse of wastewater occurs in the arid and semiarid western and southwestern states of the United States, however, the number of reuse projects is increasing in the south especially in Florida and South Carolina. Because of health and safety concerns, water reuse applications are mostly restricted to nonpotable uses such as landscape and agricultural irrigation. In a report by the National Research Council (1998), it was concluded that indirect potable reuse of reclaimed water (introducing reclaimed water to argument a potable water source before treatment) is viable. The report also stated that direct potable reuse (introducing reclaimed water directly into a water distribution system) was not practicable. Because of the concerns about potential health effects associated with the reclaimed water reuse, plans are proceeding slowly about expanding reuse beyond agricultural and landscape irrigation, groundwater recharge for repelling saltwater intrusion, and nonpotable industrial uses (e. g., boiler water and cooling water).

New Directions and Concerns

Many of the concerns mentioned in the National Research Council (NRC, 1998) report regarding potential microbial and chemical contamination of water supplies also apply to water sources that receive incidental or unplanned wastewater discharges. A number of communities use water sources that contain a significant wastewater component. Even though these sources, after treatment, meet current drinking water standards, the growing knowledge of the potential impacts of new trace contaminants raises concern. Conventional technologies for both water and wastewater treatment may be incapable of reducing the levels of trace contaminants below where they are not considered as a potential threat to public health. Therefore, new technologies that offer significantly improved levels of treatment or constituent reduction need to be tested and evaluated. Where indirect potable reuse is considered, risk assessment also becomes an important component of a water reuse investigation.

Future Trends in Technology

Technologies that are suitable for water reuse applications include membranes (pressure-driven, electrically driven, and membrane bioreactors), carbon adsorption, advanced oxidation, ion exchange, and air stripping. Membranes are most significant developments as new products are now available for a number of treatment applications. Membranes had been limited previously to desalination, but they are being tested increasingly for wastewater applications to produce high-quality treated effluent suitable for reclaimation. Increased levels of contaminant removal not only enhance the product for reuse but also lessen health risks. As indirect potable reuse intensifies to augment existing water supplies, membranes are expected to be one of the predominant treatment technologies.

Vocabulary

arid	干旱的	adsorption	吸附
semiarid	半干旱的	desalination	脱盐,脱盐作用
nonpotable	非饮用的,不适于饮用的	predominant	主要的,主导的

Reading Material B

Anaerobic Sludge Blanket Processes

One of the most notable developments in anaerobic treatment process technology was the up-flow anaerobic sludge blanket (UASB) reactor in the late 1970s in the Netherlands by Lettinga and his coworker (Lettinga an Vinken, 1980; Letting et al., 1980). The principal types of anaerobic sludge blanket processes include 1) the original UASB process and modification of the original design, 2) the anaerobic baffled reactor (ABR), and 3) the anaerobic migrating blanket reactor (AMBR). The ABR process was developed by McCarty and coworkers at Stanford University in the early 1980s (Bachmann et al., 1985). Work in the late 1990s at Iowa State University has led to the development of the AMBR process (Angenent et al., 2000). Of these sludge blanket processes, the UASB process is used most commonly, with over 500 installations treating a wide range of industrial wastewaters. A number of pilot studies have been done with the ABR, with a limited number of full-scale installations (Orozco, 1988).

The basic UASB reactor is illustrated on Fig. 27.1(a). As shown in Fig. 27.1(a), influent wastewater is distributed at the bottom of the UASB reactor and travels in an up-flow mode through the sludge blanket. Critical elements of the UASB reactor design are the influent distribution system, the gas-solids separator, and the effluent withdrawal design. Modifications to the basic UASB design include adding a settling tank (see Fig. 27.1(b)) or the use of packing material at the top of the reactor (see Fig. 27.1(c)). Both modifications are intended to provide better solids capture in the system and to prevent the loss of large amounts of the UASB reactor solids due to process upsets or changes in the UASB sludge blanket characteristics and density. The use of an external solids capture system to prevent major losses of the system biomass is recommended strongly by Speece (1996).

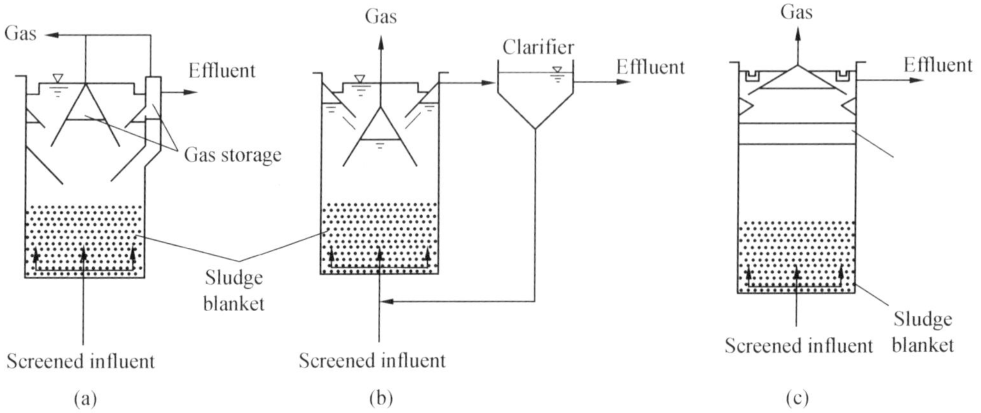

Fig. 27. 1 Schematic of the UASB and some modification: (a) original UASB, (b) UASB reacor with sedimentation tank and sludge recycle, and (c) UASB reactor with internal packing for fixed-film attached growth, placed above the sludge blanket.

The key feature of the UASB process that allows the use of high volumetric COD loadings compared to other anaerobic processes is the development of a dense granulated sludge. Because of the granulated sludge flock formation, the solids concentration can range from 50 to 100 g/L at the bottom of the reactor and 5 to 40 g/L in a more diffuse zone at the top of the UASB sludge blanket. The granulated sludge particles have a size range of 1.0 to 3.0 mm and result in excellent sludge-thickening properties with SVI values less than 20 ml/g. Several months may be required to develop the granulated sludge, and seed is often supplied from other facilities to accelerate the system startup. Variations in morphology were observed for anaerobic granulated sludge developed at 30 and 20 ℃, but both exhibited similar flock size and settling properties (Soto et al., 1997).

The development of granulated sludge solids is affected by the wastewater characteristics. Granulation is very successful with high carbohydrate or sugar wastewaters, but less so with wastewaters high in protein, resulting in a more fluffy flock instead (Thaveesri et al., 1994). Other factors affecting the development of granulated solids are maintained near 7.0, and a recommended COD:N:P ratio during startup is 300:5:1, while a lower ratio can be used during steady-state operation at 600:5:1. Control of the up-flow velocity is recommended during startup by having it high enough to wash out non-flocculent sludge.

The presence of other suspended solids in the sludge blanket can also inhibit the density and formation of granulated sludge (Lettinga and Hulshoff Pol, 1991). An explanation of the fundamental metabolic conditions associated with granular sludge formation is provided by Speece (1996) based on work by Palns et al. (1987, 1990). The explanation is as follows. The formation of dense granulated sludge flock particles is favored under conditions of near neutral pH, a plug-flow hydraulic regime, a zone of high hydrogen partial pressure, a non-limiting supply of NH_4-N, and a limited amount of the amino acid cytosine. With a high hydrogen concentration and sufficient NH_4-N, the bacteria responsible for granulation may produce other amino acids, but their synthesis is limited by the cytosine supply. Some of the excess amino acids that are produced are thought to be secreted to form extra cellular polypeptides which, in turn, will bind organisms together to form the dense pellets or flock granules.

A comprehensive review of design considerations for UASB reactors has been provided by Lettinga and Hulshoff Pol (1991). Important design considerations are 1) wastewater characteristics in terms of composition and solids content, 2) volumetric organic load, 3) up-flow velocity, 4) reactor volume, 5) physical features including the influent distribution system, and 6) gas collection system.

Vocabulary

up-flow anaerobic sludge blanket	上流式厌氧污泥床	protein	蛋白质
baffle	挡板	granulated	颗粒状的
modification	改正,修正	metabolic	新陈代谢的
fluffy	绒毛似的，蓬松的，愚蠢的	amino acid	氨基酸
		polypeptide	多肽

Unit 28 Sludge Treatment Utilization and Disposal

Source of Sludge

The first source of sludge is the suspended solids (SS) that enter the treatment plant and are partially removed in the primary settling tank or clarifier. Ordinarily about 60 percent of the SS becomes raw primary sludge, which is highly putrescent, contains pathogenic organisms, and is very wet (about 96 percent water).

The removal of BOD is basically a method of wasting energy, and secondary wastewater treatment plants are designed to reduce this high-energy material to low-energy chemicals, typically accomplished by biological means, using microorganisms that use the energy for their own life and procreation. Secondary treatment processes such as the popular activated sludge system are almost perfect systems except that the microorganisms convert too little of the high-energy organics to CO_2 and H_2O and too much of it to new organisms. Thus the system operates with an excess of these microorganisms, or waste activated sludge. As defined in the previous chapter, the mass of waste activated sludge per mass of BOD removed in secondary treatment is known as the yield, express as mass of SS produced per mass of BOD removed. Typically, the yield of waste activated sludge is 0.5 kilogram of dry solids per kilogram of BOD reduced.

Phosphorous removal processes also invariably end up with excess solids. If lime is used, the calcium carbonates and calcium hydroxyapatites are formed and must be disposed of. Aluminum sulfate similarly produces solids, in the form of aluminum hydroxides and aluminum phosphates. Even the biological process for phosphorus removed end up with solids. The use of an oxidation pond or marsh for phosphorus removal is possible only if some organics (algae, water hyacinths, fish, etc.) are periodically harvested.

Sludge Treatment

A great deal of money could be saved, and troubles averted, if sludge could be disposed of as it is drawn off the main process train. Unfortunately, the sludges have three characteristics that make such a simple solution unlikely: They are aesthetically displeasing, they are potentially harmful, and they have too much water.

The first two problems are often solved by stabilization, such as anaerobic or aerobic digestion. The third problem requires the removal of water by either thickening or dewatering. The next three sections cover the topics of stabilization, thickening, and dewatering, and then ultimate disposal of the sludge.

Sludge Stabilization

The objective of sludge stabilization is to reduce the problems associated with two detrimental characteristics—sludge odor and putrescence and the presence of pathogenic organisms. Sludge may be stabilized by use of lime, by aerobic digestion, or by anaerobic digestion.

Lime stabilization is achieved by adding lime (as $Ca(OH)_2$, or as quicklime, CaO), to the sludge and thus raising the pH to 11 or above. This significantly reduces odor and helps in the destruction of pathogens. The major disadvantage of lime stabilization is that it is temporary. With time (days) the pH drops and the sludge once again becomes putrescent.

Aerobic digestion is a logical extension of the activated sludge system. Waste activated sludge is placed in dedicated tanks, and the concentrated solids are allowed to continue their decomposition. The food for the microorganisms is available only by the destruction of other viable organisms and both total and volatile solids are thereby reduced. However, aerobically digested sludges are more difficult to dewater than anaerobic sludges and are not as effective in the reduction of pathogens as anaerobic digestion. The biochemistry of anaerobic digestion is a staged process: Solution of organic compounds by a large and hearty group of anaerobic microorganisms known, appropriately enough, as the acid formers. The organic acids are in turn degraded further by a group of strict anaerobes called methane formers. These microorganisms are the prima donnas of wastewater treatment, becoming upset at the least change in their environment, and the success of anaerobic treatment depends on maintenance of suitable conditions for the methane formers. Since they are strict anaerobes, they are unable to function in the presence of oxygen and are very sensitive to environmental conditions like pH, temperature, and the presence of toxins. A digestion goes "sour" when the methane formers have been inhibited in some way and the acid formers keep chugging away, making more organic acids, further lower the pH and making conditions even worse for the methane formers. Curing a sick digester requires suspension of feeding and, often, massive doses of lime or other antacids.

Most treatment plants have both a primary and a secondary anaerobic digester. The primary digester is covered, heated, and mixed to increase the reaction rate. The temperature of the sludge is usually about 35 ℃(95). Secondary digesters are not mixed or heated and are used for storage of gas and for concentrating the sludge by settling. As the solids settle, the liquid supernatant is pumped back to the main plant for further treatment. The cover of the secondary digester often floats up and down, depending on the amount of gas stored. The gas is high enough in methane to be used as a fuel and in fact is usually used to heat the primary digester.

All three stabilization processes reduce the concentration of pathogenic organisms, but to varying degrees. Lime stabilization achieves a high degree of sterilization, owing to the high pH. Further, if quicklime (CaO) is used, the reaction is exothermic and the elevated temperatures assist in the destruction of pathogens. Although aerobic digestion at ambient temperature is not very effective in the destruction of pathogens, anaerobic digesters have been well studied from the standpoint of pathogen viability, since the elevated temperatures should result in substantial sterilization. Unfortunately, Salmonella typhosa organisms and many other pathogens can survive digestion, and polio viruses similarly survive with little reduction in virulence. Therefore, an anaerobic digester cannot be considered a method of sterilization.

Sludge Thickening

Sludge thickening is a process in which the solids concentration is increased and the total sludge volume is corresponding decreased, but the sludge still behaves like a liquid instead of a solid. Thickening commonly produces sludge solids concentrations in the 3 percent to 5 percent range, whereas the point as which sludge begins to have the properties of a solid is between 15 percent and 20 percent solids. Thickening also implies that the process is gravitational, using the difference between particle and fluid densities to achieve the compaction of solids.

Two types of nonmechanical thickening operations are presently in use: the gravity thickener and the flotation thickener. The latter also uses gravity to separate the solids from the liquid, but for simplicity we continue to use both descriptive terms.

Sludge Dewatering

Dewatering differs from thickening in that the sludge should behave as a solid after it has been dewatered. Dewatering is seldom used as an intermediate process unless the sludge is to be incinerated and most wastewater plants use dewatering as a final method of volume reduction before ultimate disposal.

In the United States, the usual dewatering techniques are sand beds, pressure filters, belt filters, and centrifuges. Each of these is discussed in the following text. The solids concentration of the sludge from sand drying beds can be as high as 90 percent after evaporation. Mechanical devices, however, will produce sludge ranging from 15 percent to 35 percent solids.

Utilization and Ultimate Disposal

The options for ultimate disposal of sludge are limited to air, water, and land. Strict controls on air pollution complicate incineration, although this certainly is an option. Disposal of sludges in deep water (such as oceans) is decreasing owing to adverse or unknown detrimental effects on aquatic ecology. Land disposal may be either dumping in a landfill or spreading out over land and allowing natural biodegradation to assimilate the sludge into the soil. Because of environmental and cost considerations, incineration and land disposal are presently most widely used.

When sludge is destined for disposal on land and the beneficial aspects of such disposal are emphasized, sludge is often referred to as biosolids. The sludge has nutrients (nitrogen and phosphorus), is high in organic content and, as discussed, is full of water. Thus, its potential as a soil additive is often highlighted. However, both high levels of heavy metals, such as cadmium, lead, and zinc, as well as contamination by pathogens that may survive the stabilization process, can be troublesome.

Vocabulary

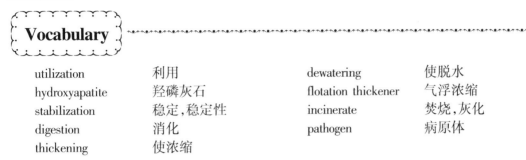

utilization	利用	dewatering	使脱水
hydroxyapatite	羟磷灰石	flotation thickener	气浮浓缩
stabilization	稳定,稳定性	incinerate	焚烧,灰化
digestion	消化	pathogen	病原体
thickening	使浓缩		

Reading Material A

Sources and Types of Solid Wastes

Knowledge of the sources and types of solid wastes, along with data on the composition and rates of generation, is basic to the design and operation of the functional elements associated with the management of solid wastes.

Sources of Solid Wastes

Sources of solid wastes are, in general, related to land use and zoning. Although any number of source classifications can be developed, the following categories have been found useful: 1) residential, 2) commercial, 3) municipal, 4) industrial, 5) open areas, 6) treatment plants, and 7) agricultural. Typical waste generation facilities, activities, or locations associated with each of these sources are presented in Tab. 28.1. The types of wastes generated, which are discussed next, are also identified.

Tab. 28.1 Typical solid waste generation facilities, activities, and locations associated with various source classifications

Source	Typical facilities activities, or locations where wastes are generated	Types of wastes
Residential	Single-family and multifamily dwellings, low-, medium-, and high-rise apartments, etc.	Food wastes, rubbish, ashes, special wastes
Commercial occasionally	Stores, restaurants, markets, office buildings, hotels, motels, print shops, auto repair shops, medicals facilities and institutions, etc.	Food wastes, rubbish, ashes, demolition and construction wastes, special wastes, hazardous wastes
Municipal *	As above	As above
Industrial demolition	Construction, fabrication, light and heavy manufacturing, refiners, chemical plants, lumbering, mining power plants, demolition, etc.	Food wastes, rubbish, ashes, and construction wastes, special wastes, hazardous wastes

Tab. 28.1(Continued)

Source	Typical facilities activities, or locations where wastes are generated	Types of wastes
Open areas	Streets, alleys, parks, vacant lots, playgrounds, beaches, highways, recreational areas, etc.	Special wastes, rubbish
Treatment plant sites principally	Water, waste water, and industrial treatment processes, etc.	Treatment plant wastes, composed of residual sludge
Agricultural	Filed and row crops, orchard, vineyard, darries, foodlots, farms, etc.	Spoiled food wastes, agricultural wastes, rubbish, hazardous wastes

* The term municipal normally is assumed to include both the residential and commercial solid wastes generated in the community.

Types of Solid Wastes

The term solid wastes is all—inclusive and encompasses all sources, types of classifications, composition, and properties. Wastes that are discharged may be of significant value in another setting, but they are of little or no value to the possessor who wants to dispose of them. To avoid confusion, the term refuse, often used interchangeably with the term solid wastes, is not used in this text.

As a basis for subsequent discussion, it will be helpful to define the various types of solid wastes that are generated (see Tab. 28.1). It is important to be aware that the definitions of solid waste terms and the classifications vary greatly in the literature. Consequently, the use of published data requires considerably care, judgment, and common sense. The following definitions are intended to serve as a guide and are not meant to be arbitrary or precise in a scientific sense.

Food Wastes Food wastes are the animal, fruit, or vegetable residues resulting from the handling, preparation, cooking, and eating of food (also called garbage). The most important characteristic of these wastes is that they are highly putrescible and will decompose rapidly, especially in warm weather. Often, decomposition will lead to the development of offensive orders. In many locations, the putrescible nature of these wastes will significantly influence the design and operation of the solid waste collection system. In addition to the amounts of food wastes generated at residences, considerable amounts are generated at cafeterias and restaurants, large institutional facilities such as hospitals and prisons, and facilities associated with the marketing of foods, including wholesales and retail stores and markets.

Rubbish Rubbish consists of combustible and noncombustible solid wastes of household, instructions, commercial activities, etc., including food wastes or other highly putrescible material. Typically, combustible rubbish consists of materials such as paper, cardboard, plastics, textiles, rubbers, leather, wood, furniture, and garden trimmings. Noncombustible rubbish consists of items such as glass, crockery, tin cans, aluminum cans, ferrous and other nonferrous metals, and dirt.

Ashes and Residues Materials remaining from the burning if wood, coal, coke, and other combustible wastes in homes, stores, institutions, and industrial and municipal facilities for purposes of heating, cooking, and disposing of combustible wastes are categorized as ashes and residues. Residues from power plants normally are not included in this category. Ashes and residues are normally composed of fine, powdery materials, cinders, and small amounts of burned and partially burned materials. Glass, crockery, and various metals are also found in the residues from municipal incinerators.

Demolition and Construction Wastes Wastes from razed buildings and other structures are classified as demolition wastes. Wastes from the construction, remodeling, and repairing of individual residences, commercial buildings, and other structures are classified as construction wastes. These wastes are often classified as rubbish. The quantities produced are difficult to estimate and variable in composition, but may include dirt, stones, concrete, bricks, plaster, lumber, shingles, and pluming, heating, and electrical parts.

Special Wastes Wastes such as street sweepings, roadside litter, litter from municipal litter containers, catch-basin debris, dead animals, and abandoned vehicles are classified as special wastes. Because it is impossible to predict where dead animals and abandoned automobiles will be found, these wastes are often identified as originating from nonspecific diffuse sources. This is in contrast to residential sources, which are also diffuse but specific in that the generation of the wastes is a recurring event.

Treatment Plant Wastes The solid and semisolid wastes from water, wastewater, and industrial waste treatment facilities are included in this classification. The specific characteristics of these materials vary, depending on the nature of the treatment process. At present, their collection is not the charge of most municipal agencies responsible for solid waste management. In the future, however, it is anticipated that their disposal will become a major factor in any solid waste management plan.

Agricultural Wastes Wastes and residues resulting from diverse agricultural activities—such as the planting and harvesting of row, field, and vine crops, the production of animals for slaughter, and the operation of feedlots—are collectively called agricultural wastes. At present, the disposal of these wastes is not the responsibility of most municipal and county solid waste management agencies. However, in many areas the disposal of animal manure has became a critical problem, especially from feedlots and dairies.

Hazardous Wastes Chemical, biological, flammable, explosive, or radioactive wastes that pose a substantial danger, immediately or over time, to human, plant, or animal life are classified as hazardous. Typically, these wastes occur as liquids, but they are often found in the form of gases, solids, or sludge. In all cases, these wastes must be handled and disposed of with great care and caution.

Vocabulary

fabrication	生产,加工	putrescible	易腐烂的
demolition	拆除,推翻	trimming	装饰物
orchard	果园	crockery	陶罐,瓦罐
encompass	围绕,包含,包围	flammable	易燃的
refuse	垃圾,废物		

Reading Material B

Sludge Dewatering

Dewatering differs from thickening in that the sludge should behave as a solid after the process. Dewatering is seldom used as an intermediate process unless the sludge is to be incinerated. Most wastewater plants use dewatering as a final method of volume reduction prior to ultimate disposal. Sludge dewatering process may be usually divided into sand drying beds and mechanical methods.

Sand Drying Beds

Sand beds have been in use for great many years and are still the most cost-effective means of dewatering when land is available and the cost of labor is not exorbitant. The beds consist of tile drains in gravel, covered by about 26 cm deep. The sludge to be dewatered is poured on the bed at about 15 cm deep. Two mechanisms combine to separate the water from the solids: seepage and evaporation. Seepage into the sand and through the tile drains, although important in the total volume of water extracted, lasts for only a few days. The sand pores are quickly clogged, and all drainage into to sand ceases. The mechanism of evaporation takes over, and this process is actually responsible for the conversion of liquid sludge to solid. In some northern areas sand beds are enclosed under greenhouses to promote evaporation as well as to prevent rain from falling into the beds.

For mixed digested sludge the usual design is to allow for three months drying time. Some engineers suggest that this period be extended to allow a sand bed to rest for a month after the sludge cake has been removed. This seems to be an effective means of increasing the drainage efficiency once the sand beds are again flooded. Raw sludge, no matter from primary or secondary settling, will not drain well on sand beds and will usually have an obnoxious odor. Hence raw sludge is seldom dried on bed.

Mechanical Dewatering Processes

Mechanical dewatering processes include belt filters, pressure filters, vacuum filters, and centrifuges. They are used in circumstances in which air-drying techniques require too much land area, odors associated with open processes are objectionable, or weather conditions do not permit their use.

Belt filters have become the most popular sludge-dewatering device in new installations. Their capital cost and operating cost are generally less than for other mechanical techniques, and they produce a sludge that can be handled as a solid. These machines are made in a variety of configurations consisting of one or more endless woven belts that pass over and around a number of cylinders. In most designs, the chemically conditioned sludge is fed to an open belt surface on which gravity drainage occurs as the belt moves forward. At the end of the gravity drainage section, provided the sludge has been properly conditioned, the solids content should be 10 percent or more. At this moisture content it behaves as a solid and can be subjected to pressure, shear, and vacuum in subsequent sections of the machine. A second belt is usually brought down toward the moving sludge at the end of the gravity zone and gradually applies pressure to the solids, squeezing out additional moisture. As the paired belts move around the rollers, their speed relative to each other varies and the sludge mass is sheared, aiding in the release of free water. The pressure gradually increases through the machine and, in some designs, a partial vacuum applied. The belts then separate, and the sludge cake is dislodged by scraping the belt or by passing it around a very small radius roller. The product can be expected to have a solids content ranging from 12 to perhaps 40 percent, and 20 percent is typical. Polymer dosages and filter loading rates are usually different for different type sludge. Problems with belt filters are primarily associated with belt tracking and belt life. Large machines may have a belt width of 3 m, and it is difficult to keep the belt on the rollers as loading conditions vary during operation. Belt life is typically between 1 000 and 2 000 hours, but may be much less if the sludge contains any large objects that can puncture the fabric.

Pressure filters consist of filter press plates. The filter support plates hold a woven filter medium against which the chemically conditioned sludge is pumped. Pumping is continued until the flow virtually ceases and the pressure is in the range of 4 to 12 MPa. The pressure is maintained for from 1 to 3 h. The frame is then opened, the plates are separated, and the cake is removed. The entire cycle can be automated, including dislodging of the dried cake with air. The solids content of the cake is often as high as 50 percent, but 30 to 40 percent is more likely. For very high solids content, addition of filter aids such as fly ash is required. Chemical conditioners for this process in the past have been largely limited to lime and ferric chloride, however, newer polymers have been applied successfully in some plants.

Vacuum filters were, for many years, the only available alternative to sludge drying beds. In new installations they are not competitive with belt filters, but, since they are used in many existing plants, a description of their use will be presented. Chemically conditioned sludge is fed to a basin in which a large rotating drum covered with a filter medium is immersed. As the drum rotates, sludge is drawn against the filter and, as it leaves the liquid, water is drawn through the medium, leaving the solids behind on the surface. Further rotation carries the solids to a removal mechanism which may involve scraping the medium with a doctor blade, passing it around a small radius roller, or both. The filter medium may be of cloth or metal. The solids content of the product may range from as little as

5 percent for unthickened waste activated sludge to as much as 40 percent for thermally conditioned sludge. Chemical conditioning is generally with lime and ferric chloride. Polymers have generally not been applied successfully on vacuum filters.

Centrifuges may be used as thickening devices for activated sludge or as a dewatering process for digested or chemically conditioned sludge. The screw conveyer or scroll rotates at a speed slightly higher than that of the bowl and thus carries the solids through the device and up the ramp to sludge cake outlet.

The variables subject to control include the bowl diameter, length, and speed; the ramp slope of length; the pool depth; the scroll speed and pitch; the feed point of sludge and chemicals; the retention time; and the sludge conditioning. Vesilind presents a complete discussion of the interrelation of these variables and their effect on solids recovery and moisture content. Recovery of fine particles and light sludge can be improved by reducing the clearance between the scroll and ramp, sometimes by pre-coating with gypsum. In general, increased solids recovery in centrifuges is associated with increased moisture content in the dewatered sludge.

Typical solids recoveries in newer centrifuges generally exceed 90 percent. Solids content depends on the source and degree of pretreatment and ranges from 12 to over 30 percent. Centrifuges are comparable to belt filters in most respects other than solids recovery. Belt filters are generally superior in that respect.

Vocabulary

cost-effective	经济的，合算的	centrifuge	离心机
mechanism	机理，机制	squeeze	挤压
seepage	渗出的量	rotating drum	转鼓
vacuum	真空	pretreatment	预处理

Unit 29 Introduction to Microbiology

What is microbiology? It is the study of microorganisms. What is a microorganism? A microorganism is something that we see only through a very high powered microscope because they are so small, we really can't see them. Microorganisms are with us all the time. Let us take an example of that. Have you ever taken the bread out of the refrigerator and leave it out and put it back for some days. Even when it's cold, mold starts growing on the bread. Where did that mold come from? It was already there. The reason the refrigerated bread grows mold is due to microorganisms in the air.

Has anybody ever seen dust floating in the air through a window? Do you think it is just dust? Most of the time it is microbes in the air. In fact, if you look in the back of the Kingsport Times, they'll say "mold spores per cubic meter". It'll give you all kinds of different spores and organisms in the air per cubic meter, and they count several thousands per cubic meter. How do they come up with that figure? A lot of scientific experimentation through the years.

No matter what it is, it can be heated on a stove. What happens when cooked food is left out in the window for a week, or a cup of coffee has been forgotten? That coffee was hot and killed all the microbes to begin with, and then it was re-inoculated, or covered with microbes again. The organisms in the air are always collecting on top of everything.

If there are thousands of organisms per cubic meter, how many are we drawing into our lungs every hour? Many, many thousands. That means that it is Nature's way that we have microorganisms. Microorganisms have been with us since the beginning of time. Even back in biblical times they knew that you could make cheese. How do you make cheese? Microorganisms.

Nothing lives without microorganisms. They are the building blocks for life. I'll give you some examples. In your intestines you have microorganisms. What do they do? They help us eat our food. The microorganisms take a little bit of our food to live, and they give us a little bit of it. They share with us. These are what we call beneficial microorganisms. Most microorganisms are beneficial. Every once in a while we'll be around someone that has the flu and we'll catch that microorganism, maybe a virus and get sick. We call that microorganism pathogenic. So we've got two classes of microorganisms, antigenic, which are the "good guys" that help us, and pathogenic, which are disease causing. Microorganisms are no more antigenic (An antigen is a chemical which stimulates certain body cells to produce a specific counteractive protein, an antibody) than any other source of foreign material, but they act as antigens more frequently, because they get into the tissues more often. Food proteins, for example, would be antigenic if injected, but they normally pass through the alimentary tract where they are digested and enter the body proper, not as proteins, but as amino acids, which are not foreign at all.

There are millions and billions of microorganisms. In fact, you can find different microorganisms and name them if you want to. You've got to provide all the information about them. Why aren't they all studied? We seem to only want to study a microorganism when it makes us sick. We invest dollars into the study of these microorganisms that make us sick because you don't want to be sick. There are new strains of microorganisms being generated every day. Microorganisms, just by their nature, change constantly. Bacteria classes are all kinds of things that change microorganisms. So the same strains we have this year may not be the same next year. What makes microorganisms do that? Take HIV, for an example. What is HIV? It's a microorganism that causes the AIDS virus. Why is everybody interested in HIV? It's a killer disease that reduces people that are very strong and healthy to very weak and susceptible people. It's a major change in lifestyle.

So let's go back to the very beginning. How did man take care of microorganisms? How did they cure food and keep it from spoiling? One of the first things they found out was that salt would preserve food and keep microorganisms from growing. All microorganisms are electrolyte specific, or salt specific. Take for example, when you're in the hospital and they're feeding you through a tube, the electrolyte has to be balanced, otherwise the shock to your system will cause you to go into shock immediately and die. This is because your salt specific isn't right. If you get higher concentrations of salt or lower concentrations of salt, you get into lots of trouble. Electrolyte was one of the first things man found out and then they found out they could cure meat by salt treating. Because the organisms that like meat are salt specific, and that salt is too concentrated for what they need to do, it helps kill microorganisms on food. It is related to osmosis.

Man also discovered that through cooking they could preserve food by getting it hot or keeping it cold. We still use that. Microorganisms, not only being salt specific, are temperature specific also. Man found out by cooking food, they could stay healthier. That helped propagate the human race. If man hadn't learned to cook food, we might still be hunting in the wilds and killing the animal, cutting it up and eating it right on the spot. What has made man different from animals? He has observed and came up with hypothesis. Man has known from a long time ago that microorganisms are temperature specific.

Pickling also keeps the microbes off food through vinegar or acidic acids. Man learned a long time ago that low pH's and high pH's preserve food. What is pH? The measure of hydrogen concentration within the solution that mimics the natural scale. Microorganisms are pH specific. The body does everything it can do to kill microbes when they get in the body. It is using all these techniques, salt concentration, pH and pickling. pH has to do with the acidity or basicness of a substance. Pickling is an example of pH, being able to preserve food with pH. Have you ever seen anybody cut the mold off cheese? Those green molds are the things that make all the amoxicillin and penicillin. Black molds do not have amoxicillin or penicillin; there are a lot of conditions that cause those black molds. If the pH of the food is low, microbes don't like to grow in our body because our pH is high and microbes don't transmit to us naturally. To get a microbe to grow that affects us, we have to have everything conducive to the growth of that microorganism.

Microbes are agar specific or food specific. Agar is a food for microorganisms. If microbes like peanut butter, they only like peanut butter. They don't adapt very readily to changes in diet. Microbes will adapt to changes in diet if you introduce a new food to them slowly and build them up to it because the enzymes they have on them are food specific. So to "teach" microbes to "eat" something they don't like, you need to put a lot of food that they like with a little bit that they don't like, this way you can eventually teach microbes to eat what they don't like.

Water is something else that affects microorganisms. Nothing lives without water. Dehydration is removing the water, and if you don't have the water, you don't have the microbes growing. That's one of the old methods for preserving food, drying it out. So change the moisture and you have decreased the possibility of these microbes growing.

Oxygen also affects microorganisms. Most organisms, other than a sulfur bacteria, require oxygen to live. It takes this energy in generating carbon dioxide. It gets its energy when it generates carbon dioxide from oxygen. The microorganism robs the oxygen from the water molecule and puts it in a carbon dioxide molecule, which on the energy scale is a lot less than the water molecule. The two basic ways that microorganisms live is by either using free oxygen or scrounging for oxygen. If you're using free oxygen, you'd expect a microorganism to eat fast. If you scrounge for oxygen, you'd expect the microorganism to eat the food slowly. So, oxygen loving microorganisms, which are aerobic, eat up materials fast like a sewage treatment plant. Non-oxygen loving microorganisms, which are anaerobic, like in a septic tank, eat up food slowly.

You have to have an absence of toxins for microorganisms. What's a toxin? Formaldehyde is one, which is hot carbon against a water molecule. All the water gases that comes from food as it's cooking has some toxins. Formaldehyde happens to be one of them. Formic acid is another one. The aldehydes are toxic to microorganisms. The formaldehyde is very toxic and when you generate that, the microorganism dies.

Back in biblical times, men would take wine skins and put grape juice and the mother of grapes which happens to be a microorganism that collects on the outside of grapes in them. Where do these microorganisms come from? They're in the air. Different grapes from different areas have different microbes on them. Different wines taste different ways. Why? Because each type of grape has different organisms in them. In the process of making wine, the key is finding the right organism and then duplicating it. Continually using that as your major organism to make wine. Some organisms give you awful tasting wine and other organisms give you sweeter tasting wine. So how does this wine form? You take the sugar that's in grape juice and put the yeast microorganism to it, then it ferments, which means it produces alcohol. The alcohol kills out the microorganisms in the wine and preserves it. If you leave that grape juice in the wine skin long enough, it becomes bitter because it's changed to vinegar. You've got to keep it cool to prevent that from occurring. So microorganisms are temperature specific.

Micronutrients are also important to microorganisms. What is a micronutrient? They are certain vitamins and minerals, some of which are found in the body, such as vitamin B-12, vitamin A, copper and other minerals. When trying to grow yeast microorganisms, you have to use brown sugar instead of white sugar. Microorganisms won't grow on white sugar because it has been bleached of all its micronutrients. You don't get all these micronutrients in white sugar or in white bread, unless you enrich it with micronutrients. Why do they bleach it in the first place? The grain. So they could make pretty white bread. People don't like that old brown bread. When they do this they have to add nutrients back to it to get the micronutrients needed. Eastman feeds locke, the micronutrients, to the sewage at their sewage treatment plant. Locke is a big user of vitamins and minerals because their substances are almost all pure substances, which doesn't grow organisms very well. So they have to feed it micronutrients.

So far, we've talked about water, electrolytes, temperature, pickling, micronutrients and toxins. Pressure is another thing that affects microorganisms. Microbes are usually pressure specific. You don't see fish down at the bottom of the ocean because they're pressure specific, or at certain pressure ranges. Microbes are also pressure specific. Microbes you find high on a mountain usually are not the same microbes you find down in the valley.

The last thing that affects microorganisms is the energy around it. As an example, if the microorganism normally grows in the sun, it has to be left in the sun. If a microorganism usually grows in the dark, it has to be left in the dark. Why? Microbes are energy specific, especially when they're not in the spore state. Spore being a state of sleep. There are microorganisms that eat PCB's. They saw something living in PCB in light concentrations.

Now all these other things have to match to make that microorganism work. Take the human body for example, what happens when you eat food? All the microbes in the air have collected on the food from the time it was on the plate to the time it got in your mouth. When you swallow the food it goes into your stomach where there is a lot of acid, which harms the microbes. Then out of the stomach into the small intestine, where we introduce bile, which is a base, and it shifts or swings all the way in the other direction. Our immune system knocks them out. Only if a lot of them get through do we have a problem, all the disease. Another person is the best vessel that handles most of the diseases that we get. Half of the people in the United States have pneumonia organisms in their body. Why doesn't the pneumonia organism overcome the body? You have resistance in the body fighting off the organism. So the most common place for us to get a disease is from other people. We can get some diseases from the soil. Legionnaire's Disease is one of those diseases. We can get some diseases from other animals, even fish. Fish can get TB and transmit it. There's another disease called Fisherman's Disease where you get a hold of the wrong fish and you can die in a couple of days from just handling it. They don't know what happens to you, it's a very tough organism that changes all the time. In controlling microorganisms, we have to either change one of the things that cause them to grow, or kill them with something that's toxic.

Vocabulary

microbiology	微生物学	electrolyte	电解，电解液
microorganism	微生物	septic tank	化粪池
HIV	人体免疫缺损病毒，艾滋病病毒	micronutrient	微量营养素
		immune system	免疫系统

Reading Material A

Process Microbiology

The anaerobic treatment process is more complex in chemistry and microbiology than aerobic treatment. A thorough understanding of both is essential by the designer and operator of anaerobic systems that are to be used successfully. This section provides the basics of anaerobic treatment microbiology and chemistry.

The consortia of microorganisms involved in the overall conversion of complex organic matter to methane begins with bacteria that hydrolyze complex organic matter such as carbohydrates, proteins, and fats into simple carbohydrates, amino acids, and fatty acids. The simple carbohydrates and acids are then utilized to obtain energy for growth by fermenting bacteria, producing organic acids and hydrogen as the dominant intermediate products. The organic acids are then partially oxidized by other fermenting bacteria, which produce additional hydrogen and acetic acid. Hydrogen and acetic acids are the main substrates used by methanogens, which convert them into methane. Hydrogen (H) is used as an electron donor, with carbon dioxide as an electron acceptor to form methane, while acetate is cleaved to form methane from the methyl group and carbon dioxide from the carboxyl group in a fermentation reaction. The complex and close community interactions of many prokaryotic organisms from two entirely different biological kingdoms Bacteria and Archaea in this widely prevalent natural process are truly amazing.

Thermodynamics and kinetics are crucial to the mixed microbial community involved in methane fermentation. Both must be understood if one is to appreciate this process in its various complexities. From an operational standpoint, the overall complexity of the process can be simplified into a few principles that can be applied readily. For instance, the process can be broken into two basic steps: 1) hydrolysis and fermentation of complex organic matter into simple organic acids and hydrogen, and 2) the conversion of the organic acids and hydrogen into methane.

The microorganisms involved in the first step grow relatively rapidly, because the fermentation reactions give a greater energy yield than the reactions that lead to methane formation. For this reason, the methanogens are more slowly growing and tend to be rate-limiting in the process. This generalization is true with domestic wastewater organic matter, municipal sludges, and most industrial wastewaters. However, with certain organic materials, for example the anaerobic decomposition of lignocellulosic materials such as grasses, agricultural crop residues, or newsprint, the hydrolysis step may be very slow and rate-limiting.

The successful start-up and operation of an anaerobic system requires that a proper balance be maintained between the hydrolytic and fermentative organisms involved in the first step and the methanogenic organisms responsible for the second step. This balance is accomplished through proper seeding, as well as through control of organic-acid production and pH during the start-up, when the microbial populations are establishing themselves. Ideally, an anaerobic reactor is seeded with digested sludge or biosolids from an active anaerobic treatment system. This kind of balanced, active seeding is necessary because of the slow double time (4 d at 35 ℃) of the critical microorganisms are required per ml of reactor volume to ensure successful operation of an anaerobic treatment system. If a seed with only 10^3 per ml of the critical organisms is available, the population would need to be increased by a factor of about 10^6. This requires about 20 doubling times, or about 80 d at 35 ℃.

At lower temperatures, the doubling time increases by a factor of about two for each 10 ℃ drop in temperature.

During reactor start-up, the operator must maintain a sufficiently small loading on the reactor so that organic acids produced by the much faster growing fermentative drop, and the methanogenic population can be killed. The crucial steps during start-up are: 1) begin with as much good anaerobic seed as possible, 2) fill the digester with this seed and water, 3) bring the system to temperature, 4) add buffering material in the form of a chemical such as sodium bicarbonate to protect against pH drop, and 5) add a small amount of organic waste sufficient to let the organic acid content from fermentation reach no more than about 2,000 to 4,000 mg/L, while keeping pH between 6.8 and 7.6. These organic acids are the food source required occurred will be evidenced through a drop in the organic acid concentration. Feeding with additional waste can then be initiated, slowly at first, until a balance is reached between the step 1 and step 2 reactions in the system. At such balance, the organic acid concentration will generally remain below 100 to 200 mg/L, depending upon the loading on the system.

Organic acid concentration and reactor pH should be determined on a daily basis to ensure that the operation of anaerobic system remains in balance. Inhibition of the biological reactions or an overload on the system with organic wastes often can be evidenced through a sudden increase in the organic acid concentration. If the buffering capacity is becoming depleted, chemical base (like bicarbonate) must then be added quickly to prevent a drop in pH from occurring, which would kill the critical methanogenic population. Thus, monitoring the organic acids and the buffer capacity provides the first line of defense for control of anaerobic systems so that the acid producers and the acid consumers achieve and maintain a proper balance.

The organic acid concentration is a key indicator of system performance. The question then is what organic acid should be measured on a routine basis, and how can this be accomplished? The key organic acids are the series of short chain fatty acids and which vary in chain length from formic acid with one carbon per mole to organic acids are the series of short chain fatty acids have been termed volatile acids because, in their unionized form, they can be distilled from boiling water. This meaning of the term volatile is different from its meaning in volatile organic compounds (VOCs), a term generally used to describe organic compounds that are readily removed from water by simple air stripping. The short-chain fatty acids cannot be removed from water by air stripping.

The volatile acids that are generally found present in highest concentrations as intermediates during start-up of an anaerobic system or during organic overload are acids found in anaerobic systems. Other nonvolatile organic acids also are formed as intermediates of waste organic degradation (e.g., lactic, pyruvic, and succinic acids) but their concentrations generally are much below those of the volatile acids and thus are of less general concern for control. The volatile acids are all quite soluble in their ionized and unionized forms and are present as dissolved species. At normal pH of operating system, they all are present for the most part in the ionized form. Typical values at 35 ℃ for the negative logarithm of the acidity constant, or pK_a(the pH at which the acids are 50 percent in the acid form and 50 percent in the ionized form), vary from a low of 3.8 for formic acid to more typical values of 4.8 for acetic and n-butyric acids and 4.9 for propionic acid.

The routine measurement of volatile acid concentration is of the greatest importance in the operation and control of anaerobic systems. Various analytical procedures can be used to measure volatile acids. Methods vary from those that require expensive instrument, such as gas chromatography and high performance liquid chromatography, to relatively inexpensive wet chemical procedures involving distillation, column chromatography, or acid/base titration. The instrumental approaches allow one to help diagnose the cause of digester failure. The wet chemical procedures generally provide information only on the total organic acids present, which can at times be used to help diagnose the cause of digester failure. The wet chemical procedures generally provide information only on the total organic acid or total volatile acid concentration present, but this is often sufficient for routine control.

A desirable analysis would be that for the active population of microorganisms present in the system, especially of the populations responsible for the critical steps of the overall process, including the methanogens. At this point, tools for such analyses are not available for routine use. However, new methods adapted from molecular biology research eventually will become available for routine operation. So far, oligonucleotide probes have been used successfully to determine the relative abundances of methanogenic populations, as well as to differentiate among the different methanogenic populations. Analyses for the more complex fatty acids that comprise bacterial cell walls has also been used to characterize bacterial populations in complex systems.

Vocabulary

carbohydrate	碳氢化合物	lignocellulosic	木质纤维
methanogen	产甲烷菌	propionic	丙酸
prokaryotic	原核的	oligonucleotide	核苷酸
Pyruvic acid	丙酮酸		

Reading Material B

Advanced Wastewater Treatment

Primary and secondary treatment remove the majority of BOD and suspended solids found in wastewaters. However, in an increasing number of cases this level of treatment has proved to be insufficient to protect the receiving waters or to provide reusable water for industrial and/or domestic recycle. Thus, additional treatment steps have been added to wastewater treatment plants to provide for further organic and solids removals or to provide for removal of nutrients and/or toxic materials.

For the purposes of this lesson, advanced wastewater treatment will be defined as: any process designed to produce an effluent of higher quality than normally achieved by secondary treatment processes or containing unit operations not normally found in secondary treatment. The above definition is intentionally very broad and encompasses almost all unit operations not commonly found in wastewater treatment today.

Types of Advanced Wastewater Treatment

Advanced wastewater treatment may be broken into three major categories by the type of process flow scheme utilized:

1. Tertiary treatment;
2. Physical-chemical treatment;
3. Combined biological-physical treatment.

Tertiary treatment may be defined as any treatment process in which unit operations are added to the flow scheme following conventional secondary treatment. Additions to conventional secondary treatment could be as simple as the addition of a filter for suspended solids removal or as complex as the addition of many unit processes for organic, suspended solids, nitrogen and phosphorous removal. Physical-chemical treatment is defined as a treatment process in which biological and physical-chemical processes are intermixed to achieve the desired effluent. Combined biological-physical-chemical treatment is differentiated from tertiary treatment in that in tertiary treatment any unit processes are added after conventional biological treatment, while in combined treatment, biological and physical-chemical treatment are mixed.

Another way to classify advanced wastewater treatment is to differentiate on the basis of desired treatment goals. Advanced wastewater treatment is used for:

1. Additional organic and suspended solids removal;
2. Removal of nitrogenous oxygen demand (NOD);
3. Nutrient removal;
4. Removal of toxic materials.

In many, if not most instances today, conventional secondary treatment gives adequate BOD and suspended solids removals. Why, then, is additional organic and suspended solids removal by advanced treatment necessary? There are a number of good answers to this question:

1. Advanced wastewater treatment plant effluents may be recycled directly or indirectly to increase the available domestic water supply.

2. Advanced wastewater treatment effluents may be used for industrial process or cooling water supplies.

3. Some receiving waters are not capable of withstanding the pollutional loads from the discharge of secondary effluents.

4. Secondary treatment does not remove as much of the organic pollution in wastewater as may be assumed.

The first three reasons for additional organic removal through advanced wastewater treatment are simple. The fourth requires some explanation. The performance of secondary treatment plants is almost always measured in terms of BOD and SS removals. A well designed and operated secondary plant will remove from 85 to 95 percent of the influent BOD and SS. However, the BOD test does not measure all of the organic material present in the wastewater. An average secondary effluent may have a BOD of 20 mg/L and a COD of 60 to 100 mg/L. The average secondary plant removes approximately 65 percent of the influent COD. Thus, when high quality effluents are required, additional organic removals must be accomplished. In addition to the organic materials remaining in most secondary effluents, there is an additional oxygen demand resulting from the nitrogen present in the wastewater.

In wastewaters, much of the nitrogen is found in the form of ammonia. When secondary treatment is used, a great deal of this ammonia is discharged in the effluent. Bacteria can utilize this ammonia as an energy source and convert ammonia to nitrite and nitrate.

$$NH_3 + O_2 + \text{Bacteria} \rightarrow NO_2 + O_2 + \text{Bacteria} \rightarrow NO_3$$

Another reason for advanced wastewater treatment may be to remove nutrients contained in discharges from secondary treatment plants. The effluents from secondary treatment plants contain both nitrogen (N) and phosphorous (P). N and P are ingredients in all fertilizers. When excess amounts of N and P are discharged, plant growth in the receiving waters may be accelerated. Algae growth may be stimulated causing blooms which are toxic to fish life as well as aesthetically unpleasing. Fixed plant growth may also be accelerated causing the eventual process of a lake becoming a swamp to be speeded up. Therefore, it has become necessary to remove nitrogen and phosphorous prior to discharge in some cases.

Toxic materials, both organic and inorganic are discharged into many sewage collections systems. When these materials are present in sufficient quantities to be toxic to bacteria, it will be necessary to remove them prior to biological treatment. In other cases, it is necessary to remove even small amounts of these materials prior to discharge to protect receiving waters or drinking water supplies. Thus, advanced wastewater treatment processes have been used in cases where conventional secondary treatment was not possible due to materials toxic to bacteria entering the plant as well as in cases where even trace amounts of toxic materials were unacceptable in plant effluents.

Nitrification

Biological nitrification may be used to prevent oxygen depletion from nitrogenous demand (NOD) in the receiving waters. Nitrification is simply the conversion of ammonia to nitrate in the treatment plant rather than in the receiving water. Nitrification may be carried out in the same tank as BOD removal or in a separate stage. Nitrification may be carried out either in activated sludge flocs or in fixed films.

Regardless of the particular scheme chosen, the same basic requirements for nitrification must be maintained:

1. Oxygen level;
2. Loading rates;
3. Solids retention time;
4. Alkalinity;
5. pH;
6. Freedom from toxic materials;
7. Temperature.

Sufficient oxygen must be available for nitrification to occur. Approximately 2.04 kilograms of dissolved oxygen are required for the conversion of 0.45 kg of ammonia to nitrate. Dissolved oxygen sufficient to satisfy the remaining BOD is also required. In activated sludge plants the mixing requirement of the basins must also be considered. Generally, dissolved oxygen levels of approximately 2-3 mg/L are recommended for nitrification. The bacteria responsible for nitrification reproduce at a much slower rate than those responsible for BOD removal. Thus, the danger always exists for the "wash out" of the nitrifying organisms. That is, unless the nitrifying bacteria reproduce at the same or greater rate than they are removed from the system (by waste sludge) then the population of bacteria will be insufficient to carry out nitrification. For this reason, nitrification systems are operated at higher return sludge rates than conventional secondary treatment. The amount of sludge to be wasted is significantly less than from a conventional activated sludge system.

Nitrification systems are sensitive to pH variation. Optimum pH has been found to be approximately 7.8 to 9.0. Reductions in nitrification have been found outside this range. Alkalinity is also destroyed during nitrification. Theoretically, 7.2 kilograms of alkalinity are destroyed in converting 1 kilogram of ammonia to nitrate. In low alkalinity wastewaters, Quick lime (CaO) or $Ca(OH)_2$ is often used to provide alkalinity and pH control.

Generally, the influent BOD to nitrification systems has not been found to effect performance. However, sufficient oxygen must be provided for the organic demand and organic shock loads must be avoided.

Biological Denitrification

Biological nitrification satisfies the nitrogenous oxygen demand by converting NH_3 to NO_3. In some applications, such as discharge into enclosed bodies of water or recycle to water supplies, nitrification may not be sufficient. When nitrogen removal is required, one of the available methods is to follow biological nitrification with biological denitrification.

Denitrification is accomplished under anaerobic or near anaerobic conditions by bacteria commonly found in wastewater. Nitrates are removed by two mechanisms: 1) Conversion of NO_3 to N_2 gas by bacterial metabolism and 2) conversion of NO_3 to nitrogen contained in cell mass which may be removed by settling.

In order for denitrification to occur, a carbon source must be available. Most commonly, methanol is used. The methanol must be added in sufficient quantity to provide for cell growth and to consume any dissolved oxygen which may be carried into the denitrification reactor.

Usually 1.36 to 1.81 pounds of methanol per pound of nitrate are required. Careful control of methanol feed is necessary to prevent waste of chemicals. In addition, if excess methanol is fed to the system, unused methanol will be carried out in the effluent causing excessive BOD.

Denitrification may be carried out in either a mixed slurry reactor or in fixed bed reactors. Denitrification filters carry out both denitrification and filtration in the same unit. Mixed slurry systems consist of a denitrification reactor, reaeration basin and clarifiers. Reaeration prior to clarification is required to free the sludge from trapped bubbles of nitrogen gas. Denitrifying bacteria grow very slowly and are extremely sensitive to temperature.

Denitrification rates have been shown to increase five-fold when the temperature is increased from 10 ℃ to 20 ℃. Thus, operating parameters such as sludge age and retention time must be varied with temperature. The pH in denitrification systems must be carefully controlled. The optimum pH is from 6.0 to 8.0.

Denitrification is a very sensitive and difficult process to operate. Little full scale operational experience is available. Constant monitoring of pH, methanol feed and temperature is essential to successful operation.

Filtration

Granular media filtration to remove those suspended and colloidal solids which are carried over from previous unit processes is a common unit process in advanced wastewater treatment. Effluents of less than 10 mg/L BOD and 5 mg/L suspended solids are not uncommon for effluents from biological treatment processes after filtration.

Gravity filters similar to rapid sand filters are sometimes used. Often a combination of filter medias, such as anthracite coal and sand are used to provide coarse to fine filtration as the water passes through the filter. The water passes through the filter media and support gravel and is then collected by the underdrain system. As filtration proceeds, the headloss through the filter increases until it reaches an unacceptable level or until solids breakthrough occurs and the effluent becomes unacceptable. When either the headloss becomes excessive or solids breakthrough occurs, the filter is backwashed.

Gravity filters are generally run at 3.66 to 6.1 m/h. Pressure filters are used to obtain filter rates up to 6 gpm per square foot. Ideally, filters are designed to have the solids in the effluent and the headloss reach their allowable levels at the same time.

Summary

Advanced wastewater treatment can be used to achieve any level of treatment desired. Advanced wastewater treatment plants in Lake Tahoe, California and Windhoek, South Africa have been achieving drinking water quality since 1968. Advanced wastewater treatment plants utilize sophisticated processes and equipment. They are relatively expensive to run and operating costs as well as effluent quality are sensitive to the quality of operation.

Vocabulary

insufficient	不够的	theoretically	理论地
nitrogenous	含氮的	underdrain	暗渠，阴沟
bloom	旺盛,茂盛	headloss	水头损失

Bibiliography

1 Metcalf, Eddy. Wastewater Engineering: Treatment and Reuse. 影印版. 北京:清华大学出版社,2003

2 J S Scott, P S Smith. Dictionary of Waste and Water Treatment. London: Butterworths Heinemann, 1981

3 Slater M J. Principles of Ion Exchange Technology. London: Butterworth Heinemann, 1991

4 Bruce E Rittmann, Perry L MacCarty. Environmental Biotechnology: Principles and Applications. 影印版. 北京:清华大学出版社,2002

5 Cleasby J L, G S Logsdon. Water Quality and Treatment: A Handbook of Community Water Supplies. 5th ed. American Water Works Association. New York: McGraw-Hill, 1999

6 Clair N Sawyer, Perry L MacCarty, Gene F Parkin. Chemistry for Environmental Engineering. 4th ed. 影印版. 北京:清华大学出版社,2000

7 W Wesley Eckenfelder, Jr. Industrial Water Pollution Control. 4th ed. 影印版. 北京:清华大学出版社,2002

8 C P Leslie Gradr, Jr. Glen T Daigger, Henry C Lim. Biological Wastewater Treatment. New York: Marcel Dekker Inc., 1999

9 Alan Scragg, Environmental Biotechnology. 影印版. 北京:世界图书出版公司, 2000

10 J Jeffrey Peirce, etc. Environmental Pollution and Control. 影印版. 北京:世界图书出版公司, 2000

11 田学达.环境科学与工程英语.北京:化学工业出版社,2002

12 杨维.水质科学与工程专业英语.北京:化学工业出版社,2002

13 朱满才,王学玲.建筑类专业英语(1).北京:中国建筑工业出版社,1997

14 傅兴海,褚羞花.建筑类专业英语(2).北京:中国建筑工业出版社,1997

15 钟理.环境工程专业外语.北京:化学工业出版社,1999

市政与环境工程系列丛书

市政工程专业英语	陈志强	18.00
环境工程土建概论	闫　波	20.00
水环境信息学	李　欣	18.80
环境化学	汪群慧	26.00
水泵与水泵站	张景成	24.00
特种废水处理技术	赵庆良	26.00
隔振降噪产品应用手册	韩润昌	30.00
污染控制微生物学	任南琪	27.00
污染控制微生物学实验	马　放	22.00
城市生态与环境保护	张宝杰	29.00
环境管理	于秀娟	18.00
水处理工程应用试验	孙丽欣	16.00
生产实习指南	刘淑彦	16.00
污水处理构筑物设计与计算	韩洪军	28.00
环境噪声控制	刘惠玲	19.80
环境科学与工程专业英语	姚　杰	12.00